08/1   11/5

AUTISM'S
FALSE
PROPHETS

Paul A. Offit, M.D.

AUTISM'S
FALSE
PROPHETS

*Bad Science, Risky Medicine,
and the Search for a Cure*

COLUMBIA UNIVERSITY PRESS

NEW YORK

Columbia University Press
Publishers Since 1893
New York   Chichester, West Sussex

Copyright © 2008 Paul A. Offit
All rights reserved
A Caravan book. For more information, visit www.caravanbooks.org.
Library of Congress Cataloging-in-Publication Data
Offit, Paul A.
    Autism's false prophets : bad science, risky medicine, and the search for
a cure / Paul A. Offit.
        p.   cm.
    Includes bibliographical references and index.
    ISBN 978-0-231-14636-4 (cloth : alk. paper)—
ISBN 978-0-231-51796-6 (electronic)
        1. Autism—Etiology.   2. Vaccination of children—United States.
3. Medical misconceptions.   4. Autism—Alternative treatment.
I. Title.   *618.92 H 11/6/08*
RJ506.A9O34 2008   *Off*
618.92'85882—dc22   2008015832

References to Internet Web sites (URLs) were accurate at the time of writing.
Neither the author nor Columbia University Press is responsible for URLs
that may have expired or changed since the manuscript was prepared.

To Kathleen Seidel, Camille Clark, Michael Fitzpatrick,
Peter Hotez, and Roy Richard Grinker:
SOME OF THE REAL HEROES—AND TRUE PROPHETS—OF THIS STORY

When religion was strong and science weak,
men mistook magic for medicine.
Now, when science is strong and religion weak,
men mistake medicine for magic.

—THOMAS SZASZ

CONTENTS

PROLOGUE

I get a lot of hate mail.

Every week people send letters and e-mails calling me "stupid," "callous," an "SOB," or "a prostitute." People ask, "How in the world can you put money before the health of someone's baby?" or "How can you sleep at night?" or "Why did you sell your soul to the devil?" They say I "don't have a conscience," am "directly responsible for the death and damage of hundreds of children," and "have blood on [my] hands." They "pray that the love of Christ will one day flood [my] darkened heart." They warn that my "day of reckoning is coming."

To understand the reason for their anger, you need to understand the decisions I've made. I'll start from the beginning.

I grew up in a suburb outside of Baltimore. My father manufactured men's shirts and my mother supervised an adult education program in the city. My first decision—to become a doctor—resulted from an accident when I was five years old. While playing on a slide at nursery school, I fell from a height of twelve feet, landing on my stomach. I lay on the ground for thirty minutes before a teacher discovered me, put me on a bus, and sent me home. Because I complained that my stomach hurt, my mother took me to our doctor, Milton Markowitz. Unfortunately, Dr. Markowitz was out of town, so we saw one of his colleagues, who assured my mother nothing was wrong; my complaints, he said, were simply an attempt to gain her attention. The pain worsened.

The next evening Dr. Markowitz came to our home. After finding the pain was greatest below the ribs on my left side, he explained that I had ruptured my spleen and it had to be removed immediately. I remember going to the hospital in my Baltimore Colts pajamas (the ones with the feet attached). It was thrilling. I was getting to ride in the car late at night while my sister had to stay home. My grandfather and I sat in the back; he put his arm around me. After the surgeon examined me, agreed with Dr. Markowitz's diagnosis, and called the head nurse to staff the operating room, my grandfather asked for another opinion, but he was told there wasn't time. At surgery they found a quart of blood in my abdomen and a spleen torn in half.

I remember Milton Markowitz. I remember his visits to our home. I remember him letting me listen to my heart with his stethoscope. I remember his big black bag with all its intricate compartments, full of glass vials and tubing and medical equipment. But mostly I remember Milton Markowitz as a kind and patient man. I wanted to be like him.

My second decision—to become an infectious disease specialist—was the result of another event that occurred when I was five. I was born with club feet, meaning that both feet turned down and inward. In those days doctors treated club feet by putting casts on both legs at birth, hoping to straighten the deformity. (In pictures of me as an infant, my legs are always covered with a blanket to hide the casts.) Although this worked to straighten my left foot, it didn't work on the right one. So in one of the first operations of its kind, I had surgery on my right foot, following which I was in a chronic care hospital for three weeks. I stayed in a large, dark room with twenty other children, all of whom had polio. Because this was a polio ward, access to the hospital was restricted; my parents could visit me on Sundays only, from two to three o'clock in the afternoon. My mother, who had had an appendectomy while pregnant with my brother, was bedridden—she never visited me. I remember just lying there, my leg strung up in a cast, staring at the front door of the hospital through a window next to my bed, hoping my parents would come. I also remember how hard it was for the other children in the ward, horribly crippled and disfigured by polio. I remember seeing them as vulnerable and helpless and alone. As I got older, the image of those children remained. I wanted to protect them, to make them feel better, to champion their causes. So I became a pediatrician and later a pediatric infectious disease specialist. (This experience also provided the emotional impetus for a book I later wrote about polio and the polio vaccine.)

I went to college in Boston and medical school in Baltimore before training in pediatrics at the Children's Hospital of Pittsburgh. In my second year of residency, I was working in the emergency room when a mother brought her nine-month-old daughter in for treatment. The young woman was from the Appalachian region around Pittsburgh, blue collar, and fiercely devoted to her daughter. She described the sequence of events leading up to her visit: her daughter had had fever, vomiting, and diarrhea; she had called her doctor and had been told to give frequent sips of water containing a little sugar and salt; because of the vomiting, she couldn't get the child to hold anything down; now, despite her efforts, her daughter had stopped urinating and had become increasingly listless.

We whisked the little girl into a room set up for emergencies and found that her blood pressure was dangerously low. To replace the fluids and minerals she had lost, we tried putting a needle into a vein in her arms, then her legs. But because she was so severely dehydrated, her veins weren't visible. In desperation, we draped her head over the side of the bed, hoping to find a vein in her neck. Again, no luck. We paged a surgeon to cut into her skin to find a deeper vein. But it would be a few minutes until he got there and her blood pressure was dropping. We didn't have time to wait. So we did something I had never seen before. The doctor in charge asked for a thick, hollow metal needle and dug it into the bone below her knee. He was hoping that fluids poured into the child's bone marrow would eventually enter her bloodstream, restore her blood pressure, and save her life. It didn't work. By the time the surgeon arrived, the child was dead. The mother was waiting outside the room. No moment is worse than this next one. We opened the door and explained to her that her nine-month-old daughter was dead because we had failed to save her. She walked into the room, knelt beside the bed, and held her daughter's hand. For the longest time we all stood there, motionless, fighting tears. The doctors in the room were all in their late twenties. None of us had children; we couldn't imagine what it would be like to lose one. (I still can't imagine this.) We could only stand by and put our hands on her shoulder and tell her how very sorry we were.

An autopsy performed several days later revealed the child's diagnosis: rotavirus. Although I had learned about rotavirus in medical school and knew that it was a common intestinal virus that caused dehydration, I didn't realize it killed children in the United States. This single experience would determine my area of research for the next twenty-five years.

Two years later, I began a fellowship in pediatric infectious diseases at the Children's Hospital of Philadelphia. My mentor was Dr. Stanley Plotkin, the inventor of the rubella (German measles) vaccine. Plotkin explained that he, along with coworker Fred Clark, had started a program to develop a rotavirus vaccine and asked if I would be interested in joining. I jumped at the chance. Working directly with Dr. Clark, we developed an animal model to study the disease. We found if we inoculated rotaviruses into the mouths of baby mice, they would get diarrhea. For the next ten years we worked to determine the parts of rotavirus that made mice sick and the parts that evoked a protective immune response. Then we constructed a series of combination rotaviruses—made from cow and human strains—that protected mice without causing disease. We were confident we had made a vaccine. At this point, Dr. Plotkin explained that pharmaceutical companies wouldn't make our vaccine if the technology wasn't protected, so we patented our vaccine viruses. (For those who hate me, this was my first crime.)

After Dr. Plotkin left Children's Hospital, Dr. Clark and I presented our ideas for a rotavirus vaccine to several pharmaceutical companies. All were interested, but we were most impressed with the scientists at Merck. Within two years our hospital entered into an agreement with Merck to license our combination rotaviruses, and for the next sixteen years Merck tested our rotavirus vaccine in progressively larger numbers of children. The final test, which included more than 70,000 infants in twelve countries, showed that the vaccine was safe and effective. In February 2006, the Food and Drug Administration (FDA) licensed our rotavirus vaccine, and the Centers for Disease Control and Prevention (CDC) recommended it for all infants. Rotavirus causes the hospitalization of tens of thousands of children in the United States and the death of about sixty every year; in the developing world, it kills 2,000 children every day. With widespread use of the vaccine, all of this suffering and death should decrease dramatically.

Although most of my hate mail mentions my work with Merck on a rotavirus vaccine, that alone doesn't explain why some people hate me. A lot of people work with pharmaceutical companies and don't get hate mail. I suspect that if I had simply continued my career in research and stayed out of the public's view, I would have escaped notice. But a series of events at our hospital in the early 1990s led to what some perceive as my second crime.

In 1991, a measles epidemic swept through Philadelphia. The outbreak centered on a religious group in the city that chose not to vaccinate its children. Seven children in that group died of measles, three in our hospital. Then the virus spread to the surrounding community, killing two more children, both of whom were too young to have gotten the measles vaccine. Because modern medicine is often incapable of preventing diseases, it's enormously frustrating. But the measles vaccine has been around for more than thirty years. It works and it's safe. Still, these parents had chosen not to protect their children.

During the next ten years, I saw several children come into our hospital with pneumonia caused by whooping cough, or severe skin infections caused by chickenpox, or meningitis caused by the bacterium *Haemophilus influenzae* type b (Hib), because their parents had chosen not to vaccinate them. When I asked why they had made that choice, they said vaccines were too dangerous: the whooping cough vaccine caused brain damage, the chickenpox vaccine caused paralysis, and the Hib vaccine caused diabetes. They had gotten their information from reports on television or the radio, from articles in newspapers and magazines, or, most commonly, from the Internet. So, in October 2000, we started the Vaccine Education Center at the Children's Hospital of Philadelphia, hoping to counter this misinformation. Within a couple of years, I was frequently quoted by the media trying to reassure parents that their concerns about vaccines were often ill-founded or had been disproved.

Although I received some hate mail for these efforts at reassurance, nothing matched what happened after the media started to carry the story that vaccines caused autism. Since the late 1990s, many studies have shown that the rates of autism are the same in vaccinated and unvaccinated children. The CDC, the American Academy of Pediatrics, and the Institute of Medicine have all issued statements supporting these studies. So the notion that vaccines cause autism isn't a medical controversy. But when I appeared on television and was quoted in newspaper and magazine articles saying that vaccines didn't cause autism, my life changed. During a congressional hearing chaired by Indiana congressman Dan Burton to investigate the cause of autism, John Tierney, a congressman from Massachusetts, asked if I had vaccinated my own children. I said I had, stating their names and ages. At the next break, a member of Tierney's staff came up to me, grabbed my arm, and pulled me aside. "Never," he said, breathlessly, "never mention the names of your own children in front of a group like this."

After I appeared on MSNBC, an extreme anti-vaccine activist called our home; later, our eleven-year-old daughter asked whether I thought anyone would ever hurt me. While I was on a federal advisory committee to the CDC—one that had made recommendations about the use of the mercury-containing preservative thimerosal in vaccines—I got a death threat. A man from Seattle wrote, "I will hang you by your neck until you are dead!" I called the CDC, which sent the e-mail to the Department of Justice, which sent it to the FBI. The threat was deemed credible, and for the next few years an armed guard was placed at the back of advisory committee meetings; for the first few months, he followed me to and from lunch, a gun hanging at his side. The mail room at my hospital regularly checks my mail for suspicious letters and packages. In June 2006, I had to walk through a rally by anti-vaccine protesters at the CDC. People shouted at me. One put a megaphone in my ear, calling me the devil. Another carried a placard with the word *Terrorist* in big red letters under a picture of me. Just before I emerged from the crowd, a man dressed in a prisoner's uniform grabbed my jacket and pulled me toward him. I don't think he wanted to hurt me; he was just excited to be close to the personification of such evil. I put my hands up in the air and asked him to please let go of my coat, which he did.

It got worse. While sitting in my office, I got a phone call from a man who said that he and I shared the same concerns. We both wanted what was best for our children. He wanted what was best for his son, giving his name and age. And he presumed I wanted what was best for my children, giving their names and ages and where they went to school. His implication was clear. He knew where my children went to school. Then he hung up.

Some people who believe vaccines cause autism hate me because they think I'm in the pocket of the pharmaceutical industry, that I say vaccines are safe because I am paid to do it. To them, it is logical that I would spend twenty-five years working on a rotavirus vaccine—a vaccine that has the chance of saving hundreds of thousands of lives every year—so that I could lie about vaccine safety and hurt children. But the reason I say vaccines don't cause autism is that they don't. I say this because the false alarm about vaccines and autism continues to harm a lot of children—harm from not getting needed vaccines, harm from potentially danger-ous treatments to eliminate mercury, and harm from therapies as absurd as testosterone ablation and electric shock. I say this be-cause the feared vaccine–autism link, which has now been dis-proved, diverts research dollars from more promising leads. I say this because I care about children with autism.

I'm not alone in this. Many parents of autistic children are angry that the media and Congress rarely talk about autism with-out blaming vaccines. And although I am certainly a target of some parents' anger, I simply represent the other side. A special kind of venom is reserved for parents of autistic children who don't believe that vaccines are at fault and actively, vigorously, and relentlessly oppose those who do. You will come to know some of these parents—the real heroes of this story—in the pages that follow.

INTRODUCTION

In 1916, polio became an American disease. In New York City alone, in one summer the virus paralyzed 10,000 people and killed 2,000. No one knew what was causing it. People blamed fish, fleas, rats, cats, horses, mosquitoes, chickens, shark vapors, pasteurized milk, wireless electricity, radio waves, tobacco smoke, automobile exhaust, doctors' beards, organ-grinders' monkeys, and toxic gases from Europe. They blamed parents for tickling their children. They blamed tarantulas for injecting poisons into bananas. Isolated, frightened, and desperate for a cure, New Yorkers tried everything. They swallowed catnip, skullcap, lady's slipper, earthworm oil, blackberry brandy, sassafras, and alcohol. Following a rumor about the curative powers of ox blood, they showed up at East Side slaughterhouses with buckets. When ox blood didn't work, they drank blood from frogs, snakes, and horses. They hung charms around their necks made of wood shavings, pepper, garlic, camphor, and onions.

Doctors offered treatments that were equally absurd. They injected adrenaline or fresh human saliva into the spines of infected children, or they took spinal fluid from infected people and injected it back under the skin. One physician, George Retan, inserted a thick, hollow needle into children's backs, drained large quantities of spinal fluid, and later infused gallons of salt water. Retan believed he was washing poisons out of the nervous system and claimed dramatic results. When it became clear that his technique was more likely to kill children than help them, Retan, who charged dearly for his therapy, was exposed as a charlatan and a fake.

Today, we see George Retan's polio cure as ill conceived and barbaric. But that's only because we know what causes polio and how to prevent it. Unfortunately, some diseases—like polio at the turn of the century—remain poorly understood. Despite advances in genetics, biochemistry, immunology, and physiology, we still don't know exactly what causes them or how to cure them. Multiple sclerosis, diabetes, and Alzheimer's disease are a few examples. Another example is autism. And like the parents of children with polio, parents of children with autism—driven by a genuine desire to help their children—are desperate for a cure. They too will try anything. Children with autism have been treated with medicines that kill fungi, parasites, viruses, and bacteria; placed on stringent diets free of grains and dairy products; subjected to high-temperature saunas; bathed in magnetic clay; asked to swallow digestive enzymes, cod liver oil, and activated charcoal; and injected with various combinations of vitamins, minerals, amino acids, and fatty acids. Although often expensive, difficult to administer, and of unproven benefit, all of these therapies are, for the most part, harmless.

Sadly, for children with autism, George Retan has cast a long shadow. In 1998, a researcher in London scared parents by claiming that autism was caused by the combination measles-mumps-rubella vaccine known as MMR. As a consequence, many parents refused the vaccine; hundreds of children in the United Kingdom and Ireland were hospitalized and four were killed by measles. At around the same time, a handful of parents claimed that a mercury-containing preservative in vaccines was to blame; they reasoned that if mercury caused autism, then ridding the body of it should help. One doctor in suburban Pittsburgh inadvertently killed a five-year-old autistic boy by injecting him with a chemical that bound to mercury but also caused his heart to stop beating. Doctors began to advocate other potentially dangerous therapies.

In 1916, George Retan was a hero. When other doctors just shook their heads, powerless to stop the growing polio epidemic, Retan stepped forward. To many parents, he was the only doctor who understood their frustration and had done something to help. He was the only doctor who cared. But Retan's therapy didn't help children with polio and occasionally killed them. Today, autism also has many well-intentioned doctors who offer parents hope for an immediate cure. But like Retan's, their therapies, which can cost tens of thousands of dollars, don't help and occasionally hurt those who are most vulnerable. The power and appeal of these modern-day false prophets—as well as the lawyers, journalists, and politicians who support them—are the subjects of this book.

AUTISM'S
FALSE
PROPHETS

CHAPTER 1

The Tinderbox

He walks as if he is in a shadow.

—LEO KANNER, DESCRIBING A CHILD WITH AUTISM, 1943

In 1938, Leo Kanner, a child psychiatrist working at Johns Hopkins Hospital, saw a five-year-old boy with symptoms he had never seen before; then he saw ten more children just like him. Five years later, Kanner published his observations in an article titled "Autistic Disturbances of Affective Contact." He introduced his paper with a plea: "There has come to our attention a number of children whose condition differs so markedly and uniquely from anything reported so far, that each case merits—and I hope will eventually receive—a detailed consideration of its fascinating peculiarities."

Kanner found that autistic children didn't talk much; when they did talk, they often talked to themselves. He also found that they played in a stereotypical and repetitive manner; demanded their toys and clothes remain in the same place every day; had an excellent memory for lists; and lacked imagination, choosing to interpret what was said to them concretely. Kanner's account of one boy, Donald, described them all: "He seemed to regard people as unwelcome intruders. When forced to respond, he did so briefly and returned to his absorption in things. When a hand was held out before him so that it could not be ignored, he played with it briefly as if it were a detached object. He blew out a match with an expression of satisfaction, but did not look up at the person who had lit the match. The most impressive thing is his detachment and his inaccessibility. He walks as if he is in a shadow, lives in a world of his own where he cannot be reached."

Kanner had used the word *autistic* in his article because he had been impressed by the children's self-absorption. (The word *autism* comes from the Greek *autos*, meaning "self.") The disorder hasn't changed. Sixty years after Kanner's original description, Ken Curtis, a radio personality from Catonsville, Maryland, described his experience: "Autism does not announce itself in the delivery room. When our son Morgan was born things were sort of a storybook for us. We had a girl and a boy, a mom and a dad, and life was kind of like a picnic. Slowly little drops of doubt began to fall. We wondered about the way he liked to watch Disney videos over and over, or how he would spin around and make strange noises and look at things out of the corner of his eye; the way he liked to line up his toys. Drop after drop we wondered, and we waited to see what would happen. He did not talk and most of the time he did not even seem to hear us. It is like being in the mall with your child, and you look down and you discover that he is not there anymore—that sickening feeling that you get in the pit of your stomach."

· · · ·

AT THE END OF HIS 1943 ARTICLE, LEO KANNER VENTURED A GUESS as to what caused autism. "We must assume," he said, "that these children have come into the world with an innate inability to form affective contact with people, just as other children come into the world with innate physical or intellectual handicaps." Kanner noticed that parents of autistic children had similar personality traits, describing them as "cold, bookish, formal, introverted, disdainful of frivolity, humorless, detached and highly—even excessively—rational and objective." He believed that children were born autistic; it was fate, destiny, beyond the control of parents and doctors, and with little hope for a cure.

Not everyone was as pessimistic as Leo Kanner. The first to offer a cure for autism was Bruno Bettelheim, a Viennese-born psychoanalyst. Bettelheim believed he had found the problem: bad mothers. He reasoned that such mothers, whom he called "refrigerator mothers," caused autism by treating their children coldly, freezing them out. If they were to recover, children with autism had to be taken from their homes and thawed. Supported by a grant from the Ford Foundation, Bettelheim founded the Orthogenic School on the South Side of Chicago. There, by replacing what he called mother's "black milk" with a supportive, nurturing environment, he claimed to have successfully treated forty autistic children, all with dramatic results. In 1967, Bettelheim published *The Empty Fortress*, in which he wrote: "Throughout this book I state my belief that the precipitating factor in infantile autism is the parent's wish that his child should not exist." In the late 1960s and early 1970s, Bettelheim promoted his ideas on television programs such as *The Today Show* and *The Dick Cavett Show*. But a closer look at Bettelheim's school showed his claims of success were fraudulent. Worse: his accusations caused mothers to feel guilty and ashamed.

• • • •

SINCE THE MID-1990S, THE NUMBER OF CHILDREN WITH AUTISM has increased dramatically. Now, as many as 1 in every 150 children in the United States is diagnosed with the disorder. Two phenomena likely account for the increase. First, the definition of autism has broadened to include children with milder, more subtle symptoms. During the time of Leo Kanner and Bruno Bettelheim, children with mild symptoms of autism may have been described as "quirky" or "different" or "unusual" but not autistic. Today, these children are more likely to be diagnosed with autistic spectrum disorder or Asperger's Syndrome or pervasive developmental delay. Second, in the past children with severe symptoms of autism were often considered mentally retarded. Today, as the number of children diagnosed with severe autism has increased, the number with mental retardation has decreased.

Because the diagnosis of autism now includes children with all forms of the illness, from mild to moderate to severe, it is difficult to talk about a single cause or treatment or cure. But there is one form of therapy that is embraced by most doctors. "Behavioral treatment," writes Laura Schreibman, professor of psychology and director of the Autism Research Program at the University of California at San Diego and author of *The Science and Fiction of Autism*, "is the only treatment that has been empirically demonstrated to be effective for children with autism." Behavioral therapy uses imitation, repetition, and frequent feedback to teach children appropriate behaviors. But because some children require a high number of repetitions, programs might require as many as forty hours a week. Although these programs can help, progress is typically slow and tortuous. Worse: they can be quite expensive and are often not covered by medical insurance. "We spent thousands of dollars," said Ken Curtis, "wrangled with the school system, hired lawyers, and lived in my grandmother's house to save on rent."

· · · ·

PARENTS OF SEVERELY AFFECTED AUTISTIC CHILDREN OFTEN FACE unimaginable emotional and financial stress. James Smythe, the father of an autistic boy in Carmel, Indiana, lamented, "Living with these children can be hell. They can destroy your entire home. You cannot keep anything nice around. They will ruin your rugs. They will move furniture around the room, push it over, break things, clear counters with one sweep of the arm. And they will do all of these things with no malice whatsoever."

Some children with autism bang their heads, bite and slap themselves, or pull their hair. Much less commonly, they gouge their eyes, causing detached retinas, or run headfirst into walls, causing fractured skulls, broken noses, and severe brain damage. One three-year-old child had a nonstop tantrum during a thirteen-hour flight from California to Germany. Such difficult behaviors have driven some parents to seek extreme medical therapies. Unfortunately, doctors have been all too willing to comply. One doctor in Massachusetts subjects children to painful electric shocks. "If it didn't hurt it wouldn't be effective," says Dr. Matthew Israel. "It has to hurt enough so that the student wants to avoid showing that behavior again."

Some parents, frustrated beyond reason and sanity, have killed their own children. On November 22, 2006, Ulysses Stable, a twelve-year-old boy with severe autism, was stabbed to death by his father in their Bronx apartment. After killing him, José Stable reportedly called the police and calmly said, "I've terminated the life of my autistic child." The police found the boy lying naked in the bathtub with a large wound starting under his left ear and extending to the middle of his throat. Two large, blood-stained knives and a meat cleaver were found in the kitchen.

On July 14, 2006, William Lash III shot and killed his twelve-year-old autistic son in their McLean, Virginia, home. Lash had been assistant secretary of commerce in the Bush administration from 2001 to 2005. A colleague later remarked, "I'm just stunned. He loved his son so much and he did everything for him."

On May 24, 2006, in Albany, Oregon, the parents of Christopher DeGroot set fire to their small apartment, locked the doors, and left their nineteen-year-old autistic son inside. A neighbor saw the flames and called 9-1-1. "I told them to get here fast because I knew a kid was inside," she said. The boy tried unsuccessfully to escape and later died with burns covering more than 90 percent of his body.

On May 13, 2006, Karen McCarron, a pathologist from Pekin, Illinois, suffocated her three-year-old daughter, Katie, with a plastic garbage bag. McCarron later told police, "Nothing is going to help and it's not going to make any difference. [I] just wanted to end my pain and Katie's pain."

Other parents have sought extreme therapies that resulted in their childrens' deaths. On August 22, 2003, Terrence Cottrell, an eight-year-old boy with severe autism, was taken to the Faith Temple Church in Milwaukee, Wisconsin, for an exorcism. His mother, Pat Cooper, told investigators that she had held down Terrence's feet while others had held his arms and head. Ray Hemphill, the preacher who had led the exorcism, pressed his knee into the little boy's chest. Two hours later, when Hemphill stood up, Terrence was dead. "We were just praying for him and asking God to deliver him from the spirit that he had," said Hemphill, who later explained that the service was in accordance with Matthew 12:43, which states, "When an evil spirit comes out of a man, it goes through arid places seeking rest and does not find it."

· · · ·

MANY PARENTS OF CHILDREN WITH AUTISM ARE TIRED OF THE glacial pace of medical research, tired of slogging through hours of behavioral therapy, and tired of watching children improve at rates so slow it's hard to tell if they are improving at all. They want something now, something that will immediately release them from the prison of autism. Douglas Biklen, a professor of special education at Syracuse University, was the first to provide it. Biklen was traveling in Melbourne, Australia, when he came upon a remarkable technique. "I [knew] that I had seen something incredible," he said. "Here was a means of expression for people who lacked expression. It was clear that this was revolutionary." He called it facilitated communication.

In 1990, Douglas Biklen brought facilitated communication to America. "Speaking involves muscles and control of muscles," he said. "In fact, it's a very complex motor activity. But [facilitated communication] is incredibly simple." Using facilitators, who held children's hands while guiding their fingers to letters on a keyboard, Biklen believed that autistic children could communicate. "It's as easy as teaching a person to eat," he said. On January 23, 1992, Diane Sawyer described Biklen's technique on the ABC news program *Primetime Live*: "Biklen began training adults called 'facilitators' to provide the lightest possible counterweight on a child's hand," said Sawyer, "to see if the experts could be wrong; that inside these autistic bodies there was someone who had something to say. And what came back was a babble of distinctive, intelligent, desperately eager voices, as if the prison doors had been opened and the prisoners could speak." The results were amazing. With the help of facilitators, children with autism typed out messages that filled their parents with hope:

"I am trapped in a cage and I want to get out."

"I am intelligent and educated."

"Autism held me hostage for seventeen years but not any more because now I can talk."

"I cry a lot about my disability. It makes me feel bad when I can't do work by myself."

"Am I a slave or am I free? Am I trapped or can I be seen as an easy and rational spirit? Am I in Hell or am I in Heaven?"

"I greatly fear for the ruin of earth unless humans jointly find a cure."

"I think you need to trust the person who facilitates. I feel I have done so well because my parents and teachers believe in me. I am also smart and facilitated communication has allowed me to show people."

"I am a Democrat and I think Anita Hill was telling the truth."

"I fear losing my ability to communicate. I fear once again being a clown in a world that is not a circus."

For decades parents had longed to communicate with their autistic children. Now, with facilitated communication, their hopes were realized. "[Her first facilitator] was a college student," said Jan Kochmeister, the mother of an autistic daughter, Sharisa. "I watched her twice and I couldn't believe what I was seeing. Since then, Sharisa's typed one hundred and twenty poems and nine short stories. She talks about everything that she feels." "He types and spells very well," said Jackie Smith of her son, Ronnie. "He knows things—words that I don't even know. After a year of facilitation, I know there's a bright, wonderful, smart, funny young man in there." "This past year, in October '92, a wonderful person came into [my daughter's] life who had gone through Doug Biklen's program here at Syracuse University," said Kathy Hayduke, "and she said, 'Do you know Stacy can write?' And I just cried. I couldn't believe it. I said, 'No, no, you're wrong. This is my kid. She's learned maybe six signs her whole life. This can't be true.' So one day [the facilitator] came over to [my] house and she said, 'Stacy, I know you're excited. After all these years, you must have something you want to tell Mom.' And Stacy typed out 'I love you, Mom.' "

Encouraged by the technique's success, Biklen started the Facilitated Communication Institute at Syracuse University. Parents and professionals flocked to see him. By 1993, hundreds of schools and centers for disabled children had adopted facilitated communication. Supported by public health departments in fifty states, Biklen's institute trained thousands of parents, teachers, speech pathologists, and health care workers—missionaries in the crusade against autism. Diane Sawyer called Biklen's technique "a miracle, an awakening." The CBS Evening News called it "a breakthrough."

Not everyone was impressed. Laura Schreibman found it difficult to believe that "even those individuals who tested in the severely retarded or severely autistic range could communicate with others, express deep emotions, write poetry, compose essays, engage in philosophical discussions, declare political affiliations, and advocate for better treatment and resources for people with disabilities." She wondered how severely autistic children could be "much more literate, mathematically skilled, insightful, and politically aware than all the professionals had suspected." Doris Allen, a psychologist at the Albert Einstein College of Medicine in New York, was also skeptical: "I think that some of the hype that facilitated communication has been receiving has been absolutely unprofessional, unvalidated, and irresponsible. [But] that happens with any new 'cure.' Many parents have come to me and requested that we [use facilitated communication]. I have not done it because [parents] already live on a roller coaster. I am in the position of having to put the brakes on the roller coaster, not to encourage a greater high and greater low for these people."

Parents who had watched the miracle of facilitated communication dismissed the warnings of psychologists like Schreibman and Allen; they knew what they had seen. And they knew Douglas Biklen had offered them something conventional researchers hadn't: hope.

It didn't last long. In 1994, some parents became suspicious when children were typing letters without looking at the keyboard. They wondered how children could type long paragraphs without spelling or grammatical errors. Finally, they asked the question they had been so careful to avoid: Who was pointing to the letters, the facilitator or the child? "It was about two years ago when someone mentioned that Dr. Biklen was giving some seminars on this facilitated communication," said Cathy Gherardi, the mother of an autistic son, Matthew. "The speech and language teacher at Matthew's school was there and she just couldn't wait to get back and start. [Matthew] was taking all kinds of Shakespeare literature, *Romeo and Juliet*, and he was in algebra class. The work that he was coming home with was absolutely incredible. Incredible." But Matthew wouldn't facilitate with his mother. "At that point I was trying to communicate with him at home," said Gherardi. "I'm saying, 'Gosh, if he's talking to these people, why isn't he talking to his mom?' You know, he and I have been best friends, bosom buddies. I mean I'm his life and he's mine. Surely we can be able to communicate. But I got absolutely nothing. As a matter of fact, he would take his [letter] board and put it in his closet and say, 'Finished.'"

Then facilitated communication tipped its hand. One of Matthew Gherardi's facilitators, Susan Rand, showed Cathy a message from Matthew claiming that he had been sexually abused by his father, Gerry. Rand reported Matthew's statements to the police. Gerry Gherardi, a pharmacist at a veteran's hospital, knew nothing of the allegations against him. "I got home around 9:30," he said. "I pulled into my driveway and Cathy came rushing down and started to talk to me. She immediately told me not to go into the house, that there was a warrant out for my arrest, that allegations had been made that I had sexually abused Matthew." Gherardi pleaded his innocence. But the school, social services, and police all believed that the accusations had come from Matthew. Gerry Gherardi spent the next six months living at a friend's house, recalling, "I told Cathy, 'There's got to be something wrong. It has to be happening someplace else. We have to call the Autism Society in Washington and find out if they have any literature on facilitated communication and allegations of sexual abuse.' When she called them up, they immediately sent us materials [that] showed that it was happening all over the country."

Through facilitated communication, autistic children in California, Texas, Georgia, Indiana, Oklahoma, and New York had claimed they had been sexually abused. Some parents, like Gerry Gherardi, had been forced to leave their homes, while others had been arrested and jailed. Children had been taken away. Then one child, a seventeen-year-old autistic girl named Betsy Wheaton, set in motion a series of events that put an end to facilitated communication.

According to her facilitator, Betsy had accused her father, mother, grandparents, and brother of sexually abusing her. Before the district attorney could prosecute the entire Wheaton family, he first had to determine who was doing the communicating. So he asked Howard Shane, an expert in communication from Boston Children's Hospital, to devise a simple apparatus, and Douglas Wheeler, a psychologist at the O. D. Heck Center in upstate New York, to test it. "We just had a table which was split down the middle," said Wheeler, a believer in facilitated communication. "The facilitator could see down one side and the student could see down [his or her] side. They couldn't see each other's sides."

One of the first children tested was Betsy Wheaton. When Betsy and her facilitator were both shown a picture of a key, Betsy typed the word *key*. But when Betsy was shown a picture of a cup and her facilitator was shown a picture of a hat, Betsy typed the word *hat*. Clearly, the facilitator was subconsciously doing the typing. Wheeler tested more children, but the results were the same. "The outcome was quite dramatic," said Wheeler. "We had zero correct responses for the students." Ray Paglieri, the director of the autism program at the O. D. Heck Center, said, "We literally didn't get one correct response. I mean, it was unbelievable given our prior belief systems about the whole thing." During the next few years, twelve studies performed in three countries showed identical results. "It was devastating to watch," said Phil Worden, Betsy's legal guardian. "Because what you saw was that the words being typed out were the words the facilitator had seen. It was just so clear and unmistakable. I was sitting there watching this and saying, 'My God, it's really true. This stuff is bogus.'"

Doug Wheeler knew that the bubble had burst. He also knew how hard it was going to be for the thousands of facilitators across the country to accept his findings. "There were 180 trials where valid communication could have been demonstrated and none did," said Wheeler. "We had overwhelming evidence for facilitator control. It began to dawn on us that the impact on the facilitators was going to be traumatic. [Facilitated communication] had become an essential part of their belief system, an essential part of their personality. People would use phrases like, 'Facilitated communication is my whole life.' These people were dedicated. They spent their own money doing training. We knew that [facilitator control] was unconscious. We knew these people had no idea they were controlling it."

Marian Pitsas, a speech pathologist and facilitator at the O. D. Heck Center, had participated in the first study showing facilitated communication was a massive, nationwide delusion. "It was devastating to see the data just there in black and white," said Pitsas. "It was mind-boggling. There was no arguing it. To see the look on Doug [Wheeler's] face, someone who I worked with, whose opinion I trust. It was devastating. I wished the floor would open up." Pitsas went back to the children to whom she had given a voice, realizing now that the communication she thought had been between her and her patient had been between her and herself. "What bothered me even more than that was thinking of the parents," said Pitsas. "We had given them false hope and we had to tell them that it wasn't real. I went back to every individual in our program that I had used facilitated communication with and tried facilitating with them with the same equipment they had used before, whether it was a keyboard, a letter board or what, without me looking at the keyboard. All I got were a string of letters. And I didn't try it just once; I went back over several days."

Douglas Biklen, the founder of facilitated communication, refused to believe the results of the studies. "People say that it must be the teachers [unintentionally] guiding," said Biklen. "It's really not possible. Students tell us information that we don't have. They tell us what went on over the weekend [and] who their siblings are. So we have all of these instances where children are telling us things that we just had no way of knowing." Facilitated communication was a matter of trust, he said. Once that trust was broken by the rigor of a study, it didn't work. "I think that the test has severe problems," said Biklen. "You're putting people in what might be described as a confrontational situation. That is, they're being asked to prove themselves. Confidence appears to be a critical element in the method. If people are anxious, they may, in fact, freeze up in their ability to respond. They may lose confidence. They may feel inadequate." Biklen claimed his method worked only when it wasn't being tested. Morley Safer, interviewing Biklen on the CBS news program *60 Minutes*, countered the claim that facilitated communication was untestable: "That's like saying, 'All pigs can fly, but they can only fly when we're not looking.'"

Howard Shane, the communication specialist who had designed the study, was sickened by the false promise of facilitated communication—that in the name of giving autistic children a voice, it had robbed them of one. "I think that it was hurtful and harmful," said Shane. "It deprived children of their right to independently communicate." Gerry Gherardi was sobered by his experience: "I think a lot of parents are grabbing at straws and are hoping that facilitated communication is going to be the answer for them, and I think they've been blinded by it. And I feel for these people because I can certainly understand where they're coming from." Douglas Wheeler was also circumspect: "If I had just thought about the literature on autism and thought about the studies I was familiar with, I would have known that the phenomenon of facilitated communication was illogical, that it probably couldn't exist. But I was so caught up in the emotionality of it."

Although it was revealed as a hoax, some parents still believe in the miracle of facilitated communication. "Something's happening here that's wonderful for all of us," said one mother. "And if it's a dream or a delusion, I'll stay on this narcotic."

. . . .

THE NEXT TO OFFER A LIFELINE TO PARENTS OF AUTISTIC CHILDREN was Victoria Beck, the mother of an autistic son. She was the first parent to urge doctors to stop treating the brains of autistic children with behavioral therapies and to start treating their intestines with medical therapies.

Parker Beck was born in January 1993. He seemed normal until a few months after his second birthday. "Then he completely zoned out and removed [himself] from the rest of the world," said Victoria. "He wouldn't respond to his name anymore and he had no interest in playing." Parker had facial tics, banged his head against the wall, and had only two words in his vocabulary: "chuss," for juice, and "k-k-k," for cracker. Worse: he suffered from chronic diarrhea and abdominal pain. In April 1996, when Parker Beck was three years old, Victoria took him to the University of Maryland medical center in Baltimore, hoping to find the cause of his intestinal problems. There, she was referred to Karoly Horvath, a pediatrician and gastroenterologist. Horvath sedated Parker, put a fiber-optic tube down his throat and into his intestines, and injected him intravenously with a hormone derived from pigs called secretin. Made by cells that line the small intestine, secretin causes the pancreas to secrete digestive juices. Horvath wanted to make sure that Parker's pancreas was working. He didn't anticipate what would happen next.

Ten days after the procedure, Parker's therapist shouted for Victoria to come downstairs quickly. "Victoria, you've got to see this," she said. Parker was pointing to pictures of his mother and father saying, "Mommy" and "Daddy." "I was stunned," recalled Victoria. "That night we picked him up and took him around the house and said 'Parker, say table.' He said 'table.' We said, 'Say candle.' He said 'candle.' 'Say chair,' and he said 'chair.' It took him a while to get the words off his tongue, but it was phenomenal." Victoria Beck believed that secretin had inadvertently treated her son's autism. Three months later, the progress stopped. "I panicked," recalled Victoria. "I was convinced this drug was helping and I couldn't get my hands on it."

Beck called the Autism Research Institute in San Diego and spoke to its founder, Bernard Rimland. Rimland had been instrumental in disproving Bruno Bettelheim's theory of the "refrigerator mother." Now, weary of the slow pace of research, he was interested in helping parents find their own cures for autism. He asked other parents if they could help Victoria Beck. One, Kenneth Sokolski, an assistant professor at the University of California at Irvine, thought he could. After watching videotapes of Parker before and after secretin therapy, Sokolski was intrigued. He persuaded a gastroenterologist in California to give secretin to his son, Aaron. "You couldn't get [Aaron] to look at you at all," said Sokolski. But after just one dose, "He looked right in the therapist's eyes."

Now two children had apparently benefited from secretin therapy. Victoria Beck called Karoly Horvath and told him about Aaron Sokolski. Horvath later published the stories of Sokolski, Beck, and one other child in a medical journal. To alert other parents that secretin might be a cure for autism, Rimland organized a conference through his Autism Research Institute. There,

Victoria Beck spoke to more than 1,000 parents. "There wasn't a dry eye in the place," recalled Rimland.

Word of the secretin miracle spread. On October 7, 1998, Jane Pauley from *Dateline NBC* told the story of Parker Beck. "It's one of medicine's most profound mysteries," she began. "But tonight a development some hail as a breakthrough may literally break the silence of autism. It's called secretin therapy and it's making a remarkable difference for some autistic children. Equally remarkable is where it came from: not medical researchers, not the federal government, but one mother and father who did what most any parent would do—they refused to give up." The demand for secretin soared. Within months, more than 2,000 parents had tried secretin on their children. Bernard Rimland called secretin "the most important development in the history of autism." But only one company, Sweden's Ferring Group, made it; and it couldn't make it fast enough. Pharmacies waited months for orders to be filled, and many parents, increasingly desperate to acquire the drug, became victims of price gouging and fraud. One Dallas doctor paid $800 for four secretin infusions and charged $8,000 to administer them. And many parents bought an oral form of secretin through the Internet later found to contain only glycerin and water.

To solve the problem of secretin supply, Walter Herlihy, a biochemist from Needham, Massachusetts, stepped forward. The father of a seven-year-old autistic girl, Herlihy too had experienced the miracle of secretin and was anxious to help other parents. He was in a unique position to do so. Herlihy was the president and chief executive officer of Repligen, a small biotechnology company that made human secretin synthetically—a purer, more natural form of the drug than the one derived from pig intestines. Herlihy hoped the Food and Drug Administration (FDA) would allow him to market human secretin for the treatment of autism. But first he had to prove that it worked.

In 1999, autism researcher Adrian Sandler and his colleagues from the University of North Carolina tested Walter Herlihy's human secretin. The group enrolled fifty-six children in the study;

half were injected with a single dose of human secretin and half were injected with salt water. For the next four weeks children were evaluated by parents, teachers, and clinicians. No one evaluating the children knew whether they had received secretin or salt water. The results were surprising. Children who had received secretin had a significant improvement in their behavioral and communication skills. But children who had received salt water also had a dramatic improvement—even greater than that for the secretin group. Sandler's study showed that secretin was no better than salt water at treating autism, but it also showed just how much parents wanted to see their children get better after an intravenous injection of a medicine they thought might help. On December 9, 1999, Adrian Sandler published the results of his study in the *New England Journal of Medicine*. When parents who had participated in the study were told that secretin didn't work, 69 percent said that they still wanted to give it to their children.

Between 1999 and 2002, investigators from three continents performed fifteen more studies of secretin. They injected autistic children with multiple doses of secretin, watched them for longer periods of time after receiving the drug, and included those with or without bowel problems. The results were the same. Not one study showed that secretin was effective in treating autism. But some parents refused to accept the outcome of these studies, preferring to believe the dramatic stories of Parker Beck and Aaron Sokolski instead. Michael Fitzpatrick, a British physician whose son has autism, has written many articles and a book on the subject. "Parents are liable to [believe in alternative] treatments for autism for a number of reasons," said Fitzpatrick. "First, the behavior of autistic children may fluctuate for no apparent reason. Second, there is a tendency for behavior to improve over time, particularly in the age range of six to ten after a more difficult period between two and five. Third, it is difficult to isolate the effect of any particular treatment from other factors that may influence behavior: a child may improve for a range of reasons not connected to the treatment. But when you have invested money, time, energy and, above all, hope in a particular treatment, it is natural to want to attribute any improvement to that treatment."

When the dust settled, Fred Volkmar, an autism researcher at Yale University School of Medicine, added a postscript to the secretin story. Echoing the words of other researchers, Volkmar wrote: "The extensive media attention [on secretin] when substantive supporting data were absent was clearly premature and unfortunate. What makes an interesting television program may not, of course, be the same as what makes good science."

Debunked, secretin is now rarely used by parents seeking a cure for autism.

· · · ·

ALTHOUGH THE FALSE PROMISES OF FACILITATED COMMUNICATION and secretin squandered research funding and drained parents' resources, these therapies weren't harmful. Unfortunately, events soon to unfold in England would dramatically raise the stakes. Parents of autistic children were about to lose much more than time and money.

CHAPTER 2

Lighting the Fuse

The trouble with the world is not that people know too little,
it's that they know so many things that just aren't so.

—MARK TWAIN

On February 28, 1998, Andrew Wakefield, a gastroenterologist working at London's Royal Free Hospital, held a press
conference. He believed he had found the cause of autism. His
findings would be published later that day in the *Lancet*, Britain's oldest and most prestigious medical journal.

Wakefield was an intriguing character. Tall, square-jawed, and
soft-spoken, with intense blue eyes and the physique of a rugby
player, he had once been a golden boy in the medical world. The
son of two doctors—his mother a general practitioner, his father
a neurologist—Wakefield had trained in Toronto before returning
to the Royal Free, one of London's most prestigious hospitals.

Before turning his attention to autism, Wakefield had studied
Crohn's disease, a chronic inflammatory disorder of the gastro-
intestinal tract, and found that it was caused by decreased blood
flow to the intestine. Researchers had spent decades trying to
find the cause of Crohn's. Now, Andrew Wakefield, only
thirty-two years old and just starting a career in academic medicine, had figured it out. Richard Horton, who would later be

named the editor-in-chief of the *Lancet*, was at the Royal Free Hospital when Wakefield made his discovery. "I was in a different department to Wakefield," recalled Horton, "but close enough to see the sensation the work caused. Research in the Royal Free's Academic Department of Medicine was largely moribund at the time I was there. [But] Wakefield brought a sudden sense of excitement. He was a committed, engaging, and charismatic clinician and scientist. The department felt alive again. He asked big questions about diseases and his ambition often brought quick and impressive results."

Wakefield knew that to help Crohn's victims, he had to find what was blocking the tiny blood vessels leading to their intestines. In 1993, he believed he had found it. In a paper titled "Evidence of Persistent Measles Virus in Crohn's Disease," Wakefield described fifteen patients with Crohn's in whom he had found measles virus in biopsies of their intestines. Natural measles virus, reasoned Wakefield, was causing Crohn's. Then he took his research one step further. In a paper titled "Is Measles Vaccination a Risk Factor for Inflammatory Bowel Disease?" Wakefield claimed that measles virus wasn't the only cause of Crohn's; measles vaccine also caused the disease. His findings shocked the medical world. Researchers in England, Denmark, Japan, Scotland, Germany, and the United States tried to duplicate Wakefield's findings but couldn't. Study after study showed that measles infections weren't more common in people with Crohn's disease; that measles virus wasn't contained in their intestines; and that vaccinated people weren't at greater risk. In 1998, in a paper titled "Measles Virus RNA Is Not Detected in Inflammatory Bowel Disease," Wakefield admitted he had been wrong. Although his findings had unnecessarily scared the public about measles vaccine, Wakefield staked a higher ground. "Hypothesis testing and presentation of the outcome—either positive or negative—is a fundamental part of the scientific process," he said. "Accordingly, we have published studies that both do and do not support a role for measles in chronic intestinal inflammation: this is called integrity."

But Andrew Wakefield wasn't finished with the measles vaccine.

. . . .

WAKEFIELD'S MISSTEP WITH CROHN'S DISEASE DIDN'T DAMPEN his enthusiasm for medical research. When he called a press conference in February 1998 to announce a possible cause of autism, journalists were impressed by his position at the Royal Free Hospital, home to an excellent medical school, and by the fact that his findings would be published in the *Lancet*. In a room packed with reporters, Wakefield announced that his team had inserted fiber-optic scopes into eight autistic children and looked at the linings of their large intestines. What they found surprised them: the intestines were studded with lymphatic nodules, similar to those found in tonsils and adenoids. Apparently, children with autism, like people with Crohn's disease, had inflamed intestines. Then Wakefield told reporters what he believed might be causing the inflammation: the MMR vaccine. All eight autistic children with intestinal inflammation had recently received MMR.

Wakefield explained how he had first made the connection between MMR and autism. "In 1995, I was approached by parents—articulate, well-educated, and concerned—who told me the stories of their children's deterioration into autism," he said. "Their children had developed normally for the first fifteen to eighteen months of life when they received the MMR vaccination. But after a variable period the children regressed, losing speech, language, social skills, and imaginative play, descending into autism." Parents of autistic children appreciated Wakefield's attention and concern. Finally, here was a doctor who cared. "He was the first doctor who really listened to us," said Isabella Thomas. "This man was my savior. He wanted to help me where others just saw me as a mother of a damaged child, an 'inconvenience.'"

During his press conference, Wakefield explained why he believed MMR might cause autism. He proposed that after MMR vaccine was injected into the arm, the measles vaccine virus trav-

Andrew Wakefield, a British gastroenterologist, proposed that the measles-mumps-rubella (MMR) vaccine caused autism (courtesy of Getty Images).

eled to the intestine and caused infection and inflammation. Harmful proteins, now able to pass through a damaged intestine, entered the bloodstream and eventually the brain, causing autism. Wakefield hadn't identified the proteins that had caused harm, but he was confident that it was only a matter of time until he did.

Wakefield also had a solution. If parents wanted to avoid autism, they should separate MMR into three vaccines. "There is sufficient concern in my own mind for a case to be made for vaccines to be given individually at not less than one-year intervals," he said. "One more case of [autism] is too many. It's a moral issue for me and I can't support the continued use of these three vaccines given in combination until this issue has been resolved."

Unfortunately, because the pharmaceutical companies that made MMR vaccine for British children didn't offer the vaccines separately, Wakefield was effectively recommending that children not be vaccinated.

Simon Murch, the second author on the *Lancet* paper and a well-respected pediatric gastroenterologist, was, like Wakefield, excited by the finding of intestinal disease in children with autism. But Murch knew that Wakefield hadn't proven his claim against MMR; he had only raised the possibility. "This link is unproven and measles is a killing infection," said Murch. "If this precipitates a scare and immunization rates go down, as sure as night follows day, measles will return and children will die." Arie Zuckerman, the dean of the Royal Free Medical School, was also surprised by the forcefulness and surety of Wakefield's statement. "Measles vaccines are among the safest and most effective vaccines ever developed," he said, pleading with the press "not [to] dent the public confidence in immunization."

But it was too late.

. . . .

ON MARCH 1, 1998—THE DAY AFTER WAKEFIELD'S PRESS conference—the British media exploded. Headlines in *The Guardian* and the *Daily Mail* read, "Alert over Child Jabs" and "Ban Three-in-One Jab, Urge Doctors." Ironically, only several years earlier public health officials had heartily supported the "Catch-Up Campaign," a program designed to immunize as many British schoolchildren with MMR as possible. Now, a respected British scientist was claiming that instead of helping children, this immunization campaign had hurt them. In response, thousands of British parents refused the vaccine.

During the next few years, journalists wrote more than 1,500 articles about Andrew Wakefield, the MMR vaccine, and autism. So great was the fascination with Wakefield that on December 15, 2003, a ninety-minute docudrama about his brilliance and courage, *Hear the Silence*, appeared on British television. The show featured the television actor Hugh Bonneville as Wakefield and Juliet Stevenson—who one year earlier had starred in the movie *Bend It Like Beckham*—as Christine Shields, the mother of an autistic son. David Aaronovitch reviewed the film for the London newspaper *The Observer*: "*Hear the Silence* [is about] the quest of a fictional mum to discover the truth about what made her son develop autism. As she quests, she hits the usual obstacles: a husband who thinks she's mad, an unsympathetic boss; and, of course, the derision of blinkered and calloused so-called experts." After the tenth doctor has refused to believe Christine's claim, she explodes. "Something happened to him," she screams. "Look at these photos. Look at how he was. Look at how he is now." Behind graphic images of the invading virus attacking the brain—shown in dramatic colors and accompanied by eerie music—Andrew Wakefield explains to Christine what has happened to her son. "It was an interior *Jaws*," said Aaronovitch. "The MMR vaccine is coming to get our kids." In *Hear the Silence*, an evil public health official plots Wakefield's demise. "The best course of action," she says, "is to discredit the work that [Wakefield's] done so we can dismiss the conclusions. I am advised that the research is flawed." "Is it?" asks a skeptical colleague. "It is [my] opinion," she replies, "and one that we should disseminate." In the movie, Wakefield is forced to withstand tremendous pressures: his files are stolen, his phone is tapped, and he hears heavy breathing on the line. Viewers learn later who is behind the intimidation: drug companies. "It's a million-pound industry," says the mother of another autistic child. "So they'll fight dirty." But Wakefield refuses to be intimidated. Turning to his wife, he says, "I wouldn't let anything happen to you or the kids. If we let them know we're scared, then they win."

Hear the Silence had an immediate impact. More British parents stopped vaccinating their children. One, Debbie Bruce, said: "It must be really frightening to be in that position when no one is listening. I'm not sure what I'm going to do about [my daughter's] next lot of injections. We're all in the dark about what is really going on." For some, frustration turned to anger. "I was overwhelmed with rage," said Leslie Mitchell, the mother of an autistic son. "I felt it building within me and it was like nothing I'd experienced before. I knew very clearly at that moment that I had crossed over to the other side, that I was convinced my son was a cash cow for an industry that tested its products in production rather than the lab, motivated by $2 billion per year in profits, no different in its potential for corruption than any other industry." Autism now had its villains: the MMR vaccine, the companies that made it, the government officials who promoted it, and the doctors who gave it. And it had its hero: Andrew Wakefield.

In the months following Wakefield's warning, the proportion of children receiving MMR vaccine dropped from nearly 90 percent to 70 percent and, in some areas of London, to 50 percent. As a consequence, small outbreaks of measles first appeared in upper-middle-class elementary schools in London. Other outbreaks followed, first in underimmunized areas of London then in Scotland and Ireland. By 2002, hundreds of children in Ireland had fallen ill with the disease. One small hospital in Dublin admitted 100 children with pneumonia and brain swelling caused by measles, three of whom died. One child, Naomi, died when the measles virus infected her brain. "By the time we got up to the ward she was dead," recalled her mother. "They were taking all the tubes out of her. When she first got sick the nurse said it was only the measles. Only?"

In 2003, the journal *Science* published an article titled "Measles Outbreaks in a Population with Declining Vaccine Uptake." Using statistical modeling, researchers showed that immunization rates in England had dropped below the level required to prevent measles from once again becoming a common childhood infection. "If the current low level of MMR vaccine uptake persists in the UK population," they wrote, "the increasing number of unvaccinated individuals will lead to a reestablishment of endemic measles and accompanying mortality." In 2006, a thirteen-year-old boy became the first person in England to die from measles in more than a decade.

· · · ·

Wakefield wasn't alone in his belief that MMR might cause autism. Following his press conference in 1998, several scientists and physicians stepped forward to support him.

John O'Leary, a professor of pathology at Coombe Women's Hospital in Dublin, examined intestinal biopsies provided by Wakefield using polymerase chain reaction (PCR), a sensitive technique to detect small fragments of measles virus genes. The results were striking. O'Leary found measles virus in the intestines of 82 percent of autistic children but in only 7 percent of nonautistic children. He concluded, "The data confirm an association between the presence of measles virus and gut pathology in children with developmental disorder."

Hisashi Kawashima, a virologist at Tokyo Medical University in Japan, found measles virus in white blood cells taken from autistic children but not from other children.

Vijendra Singh, a biologist at Utah State University in Logan, found that not only did autistic children have higher levels of measles virus antibodies in blood and spinal fluid than normal children, but they also had antibodies directed against their own brains. He believed that the MMR vaccine had caused an immune response against the sheath that lines nerve cells. Singh had extended Wakefield's hypothesis: autoimmunity, he claimed, was at the heart of the disorder.

Kenneth Aitken, a clinical psychologist at the Royal Hospital for Sick Children in Edinburgh, Scotland, and a well-known expert in the field, reviewed the cases of more than 300 children with autism and found that at least one-third of them had regressed from normal behavior before MMR vaccination to autism after it. He concluded that autism was caused by a "new environmental factor [that] could indeed prove to be MMR vaccine."

Walter Spitzer, an emeritus professor of epidemiology at McGill University in Montreal, came to the MMR controversy with an international reputation for his work showing that oral contraceptives caused blood clots. Spitzer examined Kenneth Aitken's data and agreed MMR was causing autism. But, unlike Wakefield,

Spitzer believed that autism could occur up to one year after MMR, and worried it could be even longer.

John March, a veterinarian at the Moredun Research Institute in Scotland, also came on board. March noted that measles vaccine didn't work well in animals when combined with other vaccines. He lamented that veterinarians had apparently tested animal vaccines more thoroughly than doctors had tested MMR.

Physicians from the United States also weighed in. Marcel Kinsbourne and John Menkes, pediatric neurologists from California, reviewed Wakefield's *Lancet* paper and were convinced that at least three and possibly four of the children might have been harmed by MMR. Arthur Krigsman, a gastroenterologist from the New York University School of Medicine, looked at the guts of more than forty children with autism and found—like Wakefield—that they were inflamed. Krigsman described one thirteen-year-old boy as having "the worst case of inflammation of the colon I have ever seen through a fiber-optic scope."

Andrew Wakefield had turned the British vaccination program upside down. Thousands of parents were choosing not to vaccinate their children. But Wakefield's efforts didn't end in the United Kingdom. Next, he took his case to the United States, immediately finding a powerful ally: Dan Burton, a Republican congressman from Indiana. With Andrew Wakefield as his star witness, Burton used his position as the head of the powerful Committee on Government Reform to investigate the case against MMR.

• • • •

DAN BURTON WAS A FIXTURE IN CONGRESS, FIRST ELECTED IN 1982 and reelected to his thirteenth term in 2006. Staunchly conservative and a born-again Christian, Burton is a self-made man and no stranger to public health issues. Twenty years earlier, in 1977, Burton had stood proudly on the steps of the state capitol in Indianapolis to announce he had won the right of Indiana citizens to use Laetrile, a drug made from apricot seeds, to treat cancer.

At the time, the nation's leading spokesperson for federal legalization of Laetrile was Jason Vale, a national arm-wrestling champion who claimed apricot seeds had cured his cancer. Subsequent studies of hundreds of patients showed that Laetrile didn't treat cancer, and worse, had side effects similar to those of cyanide poisoning. Burton and other Laetrile advocates argued these studies had been supported by a pharmaceutical industry bent on discrediting a cheap cure for cancer. (Actually they had been funded by the National Cancer Institute and the National Institutes of Health.) Laetrile never worked, was harmful, and is now completely discredited; but to many cancer victims at the time, Dan Burton was a hero. (Jason Vale was later convicted of fraud and sentenced to five years in prison for selling apricot seeds as a cure for cancer.)

Dan Burton's interest in drug regulation didn't end with Laetrile. When FDA commissioner David Kessler pushed to ban the use of the over-the-counter stimulant and weight-loss product Ephedra, Burton fought back. Although Ephedra clearly curbed appetites, it also constricted blood vessels and raised blood pressure. When Kessler moved to ban Ephedra, it had already caused at least fifteen deaths and hundreds of cases of psychosis, hallucinations, paranoia, depression, irregular heartbeats, strokes, and heart attacks. Problems with Ephedra were well known to the press. The *British Medical Journal* reported the case of a thirty-four-year-old man who, after taking Ephedra for ten days, had jumped out of a second-story window to escape imagined attackers. And Baltimore Orioles pitcher Steve Bechler died of a heart attack less than twenty-four hours after taking Ephedra. But in 1998, Burton accused the FDA of "harboring a culture of intimidation and sometimes harassment against alternative cures." In April 2004, despite Burton's vigorous objections, the FDA banned Ephedra from the marketplace.

Burton also joined the debate about the AIDS epidemic. After the AIDS virus entered the United States in the late 1970s, Burton became obsessed with the disease. He refused to eat soup at restaurants, brought his own scissors to the barbershop, and tried to introduce legislation requiring AIDS testing for everyone in the country—legislation that, not surprisingly, failed.

. . . .

HAVING TAKEN ON LAETRILE, EPHEDRA, AND AIDS, BURTON turned his attention to Andrew Wakefield and the MMR vaccine. On April 6, 2000, Burton and his Committee on Government Reform commanded the center of a large dais in the Rayburn House Office Building. Eight congressmen sat at his side, fifteen congressional staffers sat behind him, and hundreds of parents sat in the audience. Burton began the hearing with a statement. "I'm very proud of that picture," he said, pointing to a photograph of his granddaughter Alexandra and his grandson Christian, projected onto a large screen at the front of the room. "The one on the left is my granddaughter. My grandson—who you see there with his head on her shoulder—after receiving nine shots on one day, quit speaking, ran around banging his head against the wall, screaming and hollering and waving his hands, and became totally a different child. And we found out that he was autistic. He was born healthy. He was beautiful and tall. He was outgoing and talkative. He enjoyed company and going places. Then he had those shots and our lives changed and his life changed." Burton stopped, trying to compose himself. "I don't want to read all of the things that happened to Christian," he said, "because I don't believe that I could make it through it. But I can't believe that this is just a coincidence. And when people tell me that that's just a genetic problem I'm telling you that they're just nuts. That's not the way it was." Later, Burton described the importance of the hearing: "If we want to find a cure, we must first look to the cause. We must do this now, before our health and education systems are bankrupted, and before more of our nation's children are locked inside themselves with this disease." Dennis Kucinich, a congressman from Ohio and soon-to-be presidential candidate, agreed with Burton. "The problem," he said, "is what we are giving our children." Later, Burton's daugh-

ter, Danielle Burton-Sarkine, would be one of the first Americans
to file a lawsuit against drug companies, claiming that vaccines
had caused her son's autism.

After Burton was finished, it was the parents' turn to testify.
Sitting at a long table at the foot of the dais, several parents—like
Burton, angry that MMR had damaged their children—told
their stories. Shelley Reynolds said: "Right after [we] were married, Hurricane Andrew, one of the most destructive hurricanes
to ever hit the United States, slammed through Baton Rouge. We
promised each other we would never again be unprepared for such
a disaster. But six years later hurricane-force winds blew into our
home again. This time the disaster was the diagnosis of autism
for our first-born son, Liam. What had happened to our beautiful baby boy? [I] have no doubt that he developed autism as a
direct result of [a] vaccine. The pain of knowing that I inadvertently caused him harm due to blind trust in the medical community is nearly unbearable." Reynolds then pointed a finger at
the real villain: drug companies. "They obviously have a forced

*Dan Burton, a Republican congressman from Indiana, used his position as
chairman of the Committee on Government Reform to investigate the
relationship between vaccines and autism (courtesy of Getty Images).*

market," she said. "They manufacture products that are required for every child in this country. While seeking greater profits, [they] have lost sight of the medical community's original goal—to protect children from harm."

Jeanna Smith, from Denham Springs, Louisiana, said: "At sixteen months of age, Jacob received his MMR vaccine; immediately following that, [he] began exhibiting strange behaviors. I cannot bear the thought that after being so careful, I was so easily persuaded to immunize Jacob without knowing all that I should know. Jacob's countenance left when he was sixteen months old. The light behind his eyes was replaced with a blank, lost, bewildered stare."

Scott Bono, from Durham, North Carolina, was also convinced that MMR had caused his son's autism. "On August 9, 1990, Jackson [began] his journey into silence," said Bono. "That was the day he received his MMR immunization. He would not sleep that night. In the days to follow he would develop unexplained rashes and horrible constipation and diarrhea. His normally very healthy body was being ravaged by an invader. Over the next weeks he would slip away, unable to listen or speak. What was the reason for the change? It is my sincerest belief that it was that shot."

At the end of the session, the parents who had testified slowly stood up and filed out of the meeting room, each looking back at Burton, hopeful they would soon get the answers they needed. Next up were scientists who studied autism—scientists who, like the parents who had just testified, believed vaccines were the cause.

· · · ·

ANDREW WAKEFIELD, NOW WELL KNOWN TO AMERICAN PARENTS and the press, was the first to testify. Burton treated him like a celebrity. Unfortunately, Wakefield presented his evidence as if he were speaking to scientists, not congressmen and parents. He showed pictures of intestines of autistic children viewed from the end of a fiber-optic scope, intestinal cells containing small fragments of measles vaccine virus, and plastic gels containing measles virus proteins. He talked about follicular dendritic cells, ileocolonic lymphoid nodular hyperplasia, crypt abscesses, common recall antigens, molecular amplification technology, and hepatic encephalopathy, concluding, "We have a biologically plausible hypothesis, that undegraded chemicals from the gut may be getting through and impacting the rapidly developing brain during the first few years of life [causing] autism." Then he surprised the committee. He presented findings not only from the eight autistic children in his *Lancet* paper, but also from 150 children in whom he claimed MMR had caused autism in "all but four."

John O'Leary, the Irish pathologist who had identified measles virus in the intestines of Wakefield's autistic patients, was next. He too presented as if he were speaking to a group of molecular biologists. O'Leary talked about TaqMan real-time quantitative PCRs, RNA inhibition assays, low-copy viral gene detections, fusion proteins, neuraminidases, hemagglutinins, subacute sclerosing panencephalitis, nucleocapsid genes, and black signals, concluding, "I am here to say that Wakefield's hypothesis is correct."

Other scientists at Burton's hearing were quick to support Wakefield's notion that MMR caused autism, each seeing an opportunity to proffer their own therapies. Mary Megson, a pediatrician from the Medical College of Virginia, said, "The MMR vaccine depletes the body of all its stores of vitamin A. When I give [autistic children] vitamin A in the form of cod liver oil, they get well. In a few days a lot of these children look at me, focus, regain eye contact, and talk about their box of vision growing. One child's IQ score went up 105 points." Michael Goldberg, a pediatrician from Los Angeles, said, "I am here before you today to share my frontline, everyday experience with these children; experience that has overwhelmingly convinced me and my colleagues that this is a disease that can be treated." Goldberg claimed dramatic results with minerals, anti-virals, anti-fungals, anti-inflammatories, and immune-modifying agents. Vijendra Singh, the Utah State University biologist who had found unusual

levels of antibodies in the blood and spinal fluids of autistic children, suggested using steroids, intravenous gamma globulin, and
an occasionally dangerous technique called plasmapheresis, during which blood is removed and replaced with blood free from
harmful antibodies. Finally, John Upledger, an osteopath and
head of the Upledger Clinic in Palm Beach Garden, Florida, explained that vaccines had caused an abnormal flow of spinal fluid.
"It looked like autistic children were trying to knock something
loose from their heads," he said, "something that was jammed
together. The cranial rhythm or the movement inside the head
was not giving the amplitude that we were looking for." But
Upledger had a solution. "We started decompressing heads," he
said, "forehead forward, back of the head backward. We would
just sit there and hold it, a small force over a long period of time,
and ultimately the head would begin to expand. The thumb-sucking
stopped, the head-banging stopped, and the wrist-chewing
stopped. When we were finished, they were very liable to turn
around and kiss you and give you a hug. And after that they became sociable. There was close to a 100 percent response."
Upledger later explained that autism was different in Europe.
"When I went to Brussels," he said, "it was an entirely different
thing. The feel of the head, the energy patterns of the head, everything was different." Burton, who had struggled with the scientific
presentations of Wakefield and O'Leary, understood and appreciated Upledger's. Clearly excited, he said, "That was a really
good lecture. I enjoyed that and we will have some questions
about whether or not any of our health agencies have picked up
on your procedures."

Although Burton was the chairman, he wasn't the only member of the committee to determine who would testify. Henry
Waxman, a congressman from California and the ranking member on the committee, also had a say. Although Waxman had
made a career badgering pharmaceutical companies about the
price and availability of their products, he had always supported
good science and the mission of the National Institutes of Health
(NIH). Waxman chose Brent Taylor, a professor of community

medicine and child health, who, like Wakefield, worked at the Royal Free Hospital in London.

Taylor started his testimony with a bombshell. "Mr. Wakefield's and Professor O'Leary's testimony notwithstanding, the belief that MMR is the cause of autism is a false hope," said Taylor. "There is no evidence that immunizations are involved." Taylor showed his data. He found that autism rates in the United Kingdom had clearly been rising before MMR was first introduced in 1988 and that children who had received the vaccine weren't at greater risk. Burton was angry. This wasn't how he had wanted the hearing to go. But he was prepared, having been provided a rebuttal by the Canadian epidemiologist Walter Spitzer. Spitzer had told Burton that Taylor had knowingly excluded some children from his study. "In your *Lancet* paper, you omitted to mention the vaccination of children over one year of age when the vaccination was introduced," said Burton. "Do you think that is a correct analysis?" "Basically, that statement is not true, Mr. Chairman," said Taylor. "We did include [those] children in our analysis as we clearly stated." (Taylor had already made this point during his presentation but Burton hadn't heard him.) "To suggest otherwise," said Taylor, "and I suspect the suggestion comes from Mr. Wakefield, is malicious." "[Were the data] omitted from the original paper that you submitted?" Burton persisted. "No," said Taylor. "All cases were included who received MMR vaccine." Burton was stuck. He'd probably never read Taylor's paper or, for that matter, Wakefield's, Singh's, or O'Leary's. It wasn't that he wasn't interested in their studies; he was. It was just that he didn't have the scientific background necessary to fully understand them. Burton sat quietly, glaring at Taylor.

Waxman was upset by the heated exchange between Burton and Taylor. "Mr. Chairman," he said, "because I believe autism is such a serious problem, I am troubled by this hearing. This hearing was called and structured to establish a point of view, and it is the point of view of the chairman [Burton]—that is the connection between autism and vaccinations. Dr. Wakefield came out with a report in England, and the first group that examined his claims was the Medical Research Council, and they found no evidence to indicate a link between the MMR vaccine and bowel disease and autism. Then the World Health Organization looked at his study and they came up with the following statement: 'Given our view, the claims made by Dr. Wakefield and his colleagues lack scientific credibility; and this present study does not meet the requirements of establishing a causal relationship [between MMR and autism].' Now, Dr. Wakefield has testified that he has some new information. Fine. Let us get the new information out there. Let us let the epidemiologists evaluate it. Let us let the scientists explore where the real truth may be."

Waxman felt that scientific questions like "Does the MMR vaccine cause autism?" were best determined by scientists, not congressmen. "I cannot tell you what is true or not," said Waxman, "but I do not think that our chairman can tell you either." Waxman knew that much was at stake. Referring to previous Burton hearings on the Whitewater scandal involving Bill and Hillary Clinton, Waxman said, "I feel that when we had the hearings on campaign abuses by Democrats, a lot of people's reputations were ruined and I thought the hearings were unfair. But those were political. The consequences of an unfair hearing on autism connected to vaccinations can cause people to die."

The American media loved Dan Burton's hearing. The *New York Times*, CNN, *USA Today*, the *Washington Post*, and almost every major newspaper, magazine, and radio and television station in the United States reported that the MMR vaccine—used in the United States for almost thirty years—might cause autism. On November 12, 2000, seven months after Burton's autism hearing, *60 Minutes* produced a program titled "The MMR Vaccine." The segment began with correspondent Ed Bradley interviewing Dave and Mary Wildman, the parents of an autistic son, from Evans City, Pennsylvania. The boy, Bradley said, "appeared perfectly normal until just after his first birthday, when he received the MMR vaccine. Within a few weeks, according to his parents, things began to change." "He started to not look at me anymore when I would call his name," said Mary.

Henry Waxman (left), *a Democratic congressman from California, frequently sparred with Dan Burton during the Committee on Government Reform's investigation of vaccines (courtesy of Getty Images).*

"And do you know why?" asked Bradley. "Because of the MMR vaccine," she said, tears streaming down her face. "I should never have had him have that vaccine." Bradley also interviewed Andrew Wakefield. "My concern comes initially from the story the parents tell," said Wakefield. "They have a normally developing child who upon receipt of the MMR vaccine develops a complex syndrome of behavioral and developmental regression, loss of speech, loss of language, loss of acquired skills, [and] loss of socialization with siblings or peers." "Do you have children?" asked Bradley. "I have four children," Wakefield replied. "Knowing what you know now," said Bradley, "would you give them the MMR vaccine?" Wakefield leaned toward the camera, calm and certain. "No, I wouldn't," he said. "I would most certainly vaccinate them. I would give them [separate shots of] measles, mumps, and rubella vaccines."

Wakefield's claims were picked up by news agencies across the world. As a consequence, the Ministry of Health in Japan suspended its recommendation to use MMR and more than 100,000 American parents decided not to give it to their children. Michael Fitzpatrick, the British physician whose son is autistic, commented on Wakefield's impact: "Dr. Wakefield's paper appears to be a dramatic example of the butterfly effect celebrated in chaos theory, in which the flutter of tiny wings on one continent is amplified around the planet to produce a tidal wave on some distant shore."

• • • •

ANDREW WAKEFIELD HAD TAKEN HIS CASE TO THE BRITISH AND American public with dramatic results. Several well-respected scientists and clinicians had now lined up behind him. Hundreds of thousands of parents had chosen not to give MMR vaccine to their children. Even actors and sports celebrities were stepping forward to convince the public and the press that vaccines were unsafe. Wakefield and his MMR-autism theory were riding high. But in 2004, an investigative journalist in London found that Andrew Wakefield wasn't exactly what he appeared to be. In a blink, his science, his theories, and his career all came crashing down.

CHAPTER 3

The Implosion

While Galileo was a rebel, not all rebels are Galileo.

—Norman Levitt

On February 18, 2004—six years after Andrew Wakefield had published his paper in the *Lancet*—Brian Deer, an investigative reporter for the *Sunday Times* of London, called Richard Horton, the *Lancet*'s editor-in-chief. Deer said he had some shocking news about Andrew Wakefield. Horton gathered several other editors into the *Lancet*'s editorial office and waited; he had no idea what he was about to hear. The meeting lasted five hours. "The allegations made by Deer, as I saw them," recalled Horton, "were devastating."

Deer claimed that Wakefield's *Lancet* paper contained several errors. For one, Wakefield had written that his "investigations were approved by the Ethical Practices Committee." But, according to Deer, the committee had never approved Wakefield's study. Wakefield's team had put children under general anesthesia, performed spinal taps, threaded fiber-optic scopes into their intestines, taken biopsies, and collected large quantities of blood for testing. These procedures weren't trivial. Several children had difficulties with the anesthesia, and one five-year-old child was in critical condition after his colon was perforated in several places. If the Ethical Practices Committee hadn't granted its approval, then Wakefield and his coworkers had circumvented a process designed to protect children from unnecessary and potentially harmful tests—a serious charge.

There were other problems. Wakefield had written that his "study was supported by the Special Trustees of Royal Free Hampstead National Health Service Trust and the Children's Medical Charity." But he didn't mention the study's largest supporter. Two years before his paper was published, Deer claimed that Wakefield had been given $100,000 by a personal-injury lawyer named Richard Barr. At least five of the eight autistic children in Wakefield's study were clients of Barr. Deer claimed Wakefield knew that parents of the children in his study had a financial interest in finding a link between MMR and autism; if Wakefield could establish that link, these parents could successfully sue for compensation. Deer also challenged Wakefield's claim that his interest in MMR stemmed from encountering autistic children during their routine admissions to the hospital. More likely, Deer believed, lawyers had referred patients to Wakefield, who then laundered their stories into a medical publication, grossly misleading the reader. "If this claim were true," recalled Horton, "it would have meant that the selection of children who took part in the investigation had been badly biased."

• • • •

WAKEFIELD'S CONNECTION TO A CLASS-ACTION LAWSUIT AGAINST vaccine makers began in 1996—two years before his *Lancet* publication—when he met Rosemary Kessick. Kessick was concerned about what had happened to her son after he had received MMR. "He looked strange," remembered Kessick, a full-time business analyst. "He had a kind of yellow tone. He wasn't well. Then [he] developed chronic diarrhea." Fearing MMR had caused her son's illness, Kessick took him to several doctors, all of whom reassured her that it was probably just a virus. "Most of them were like, 'Oh, don't worry your little head about the MMR,'" she recalled. "[But] within weeks of the vaccination, [my son's] development slowed down, then it stopped and he regressed. Seeing what has happened to him broke our hearts." Through a parents' advocacy group, Kessick had learned about Andrew Wakefield who, after examining her son, agreed that MMR had caused his autism. "It means so much to finally be listened to," said Kessick, "and to find people prepared to stand up and say the safety of these vaccines must be investigated."

Immediately after leaving Wakefield's office, Kessick visited the office of Justice, Action, and Basic Support (JABS), an organization that warns parents about the dangers of MMR. (Jabs is the British word for shots.) JABS members believed MMR caused mental retardation, epilepsy, arthritis, and weakened immune systems; Kessick was the first parent to believe it caused autism. In 1994, two years before recruiting Andrew Wakefield to their cause, members of JABS had sued the three pharmaceutical companies that made MMR: Merck, SmithKline Beecham (now GlaxoSmithKline), and Aventis Pasteur (now Sanofi Pasteur). The personal-injury lawyer who represented them was Richard Barr, at the time a partner in the British law firm Dawbarns.

. . . .

RICHARD BARR HAD CONSIDERABLE EXPERIENCE SUING PHARMA-ceutical companies, first coming to public attention during his crusade against the anti-arthritis drug Opren. Opren had already been withdrawn from the United States, where it had been linked to more than seventy deaths and 1,000 cases of kidney and liver failure. The drug's manufacturer, Eli Lilly, eventually settled cases in Britain for $6 million. Unfortunately, because he had filed his claims after an agreed-upon deadline, Barr's clients received nothing. "My worst day as a lawyer," recalled Barr "was when the first judgments were handed down in the Opren cases. When [the judges] dismissed all my claims, it was like a firm kick in the solar plexus." Barr had also crusaded against organophosphate chemicals, claiming that farmers exposed to pesticides and soldiers exposed to nerve gases suffered depression, irritability,

difficulty concentrating, and poor sleep (Gulf War Syndrome).
Again, Barr failed to obtain compensation for his clients. During the 1990s, Barr moved from one British law firm to another;
from Dawbarns in Ipswich he went to Hodge, Jones and Allen
in Camden Town and then to Alexander Harris in Manchester,
each time taking with him his ever-growing caseload of claims
against MMR. After Rosemary Kessick walked into his office—
and after the association between MMR and autism first appeared in a British newspaper in November 1996—the number of
claims against MMR increased from hundreds to thousands.

To Richard Barr, the case against MMR was clear. "If it can
be shown objectively," he said, "that a child was developing normally prior to being vaccinated; and if there was no other event
which could account for the condition, then in all probability
the MMR vaccination has played a part in the cause of autism."
According to Brian Deer, Barr had paid Andrew Wakefield more
than $100,000 to prove it.

• • • •

AFTER HEARING DEER'S ALLEGATIONS, RICHARD HORTON ASKED
Andrew Wakefield, Simon Murch, Peter Harvey, and John Walker-
Smith—four of the thirteen authors on the *Lancet* paper—to
come to his office. Horton told them what Deer had told him,
recalling, "Simon Murch and John Walker-Smith [a senior gas-
troenterologist] were visibly shocked by this revelation." Deer
had asked Horton not to talk to the press until after he had published his story in the *Sunday Times*. But Horton, reasoning
that it was in everyone's best interest, immediately called the
BBC. On February 20, 2004, Horton appeared on BBC Television. "If we knew then what we know now," he said, "we certainly would not have published the part of the paper that
related to MMR." In the newspaper interviews that followed,
Horton said, "There were fatal conflicts of interest in this paper.
As the father of a three-year-old who had had MMR, I regret
hugely the adverse impact this paper has had."

Following Horton's appearance on BBC Television, reporters cornered Wakefield. After denying any wrongdoing, Wakefield admitted that "four, perhaps five" of the children in the *Lancet* study were clients of Richard Barr. "Was it four or five?" they asked. "Let's make it five," he said. "Were they litigants?" "Yes," replied Wakefield. "Were you being paid to help them build their case?" Again, Wakefield said yes. "Did you tell your colleagues that these children were part of the study?" "I don't recall," said Wakefield. "Did you tell the *Lancet* about these conflicts prior to the publication?" Wakefield said he hadn't. "Why not?" "I believe that this paper was conducted in good faith," said Wakefield. "It reported the findings. There was no conflict of interest." "Do you have any reasons now to change your opinion?" "No," said Wakefield, "but again it's a debate. I have no regrets." Wakefield also disputed the alleged sum of $100,000, claiming that $50,000 was closer and that he had given all of the money to a research assistant. Like Wakefield, Richard Barr pleaded his innocence. "We weren't trying to get an independent paper published under the carpet," said Barr. "I remember noting at the time that the funding acknowledgment wasn't there, but it didn't seem to be a big deal."

Simon Murch was furious that Wakefield had never told him about the money from Richard Barr. "It was a very unpleasant surprise," said Murch. "We never knew anything about the [money]. He had his own separate research fund. We were pretty angry." Murch called other members of the Royal Free team and explained what had happened. He asked whether they would be interested in lining up behind him to retract the findings of the paper. Ten of the thirteen agreed. On March 6, 2004, three weeks after Brian Deer called Richard Horton, the Royal Free Hospital team's retraction appeared in the *Lancet*: "We wish to make it clear that in this paper *no causal link was established between MMR vaccine and autism* as the data were insufficient. However, the possibility of such a link was raised and consequent events have had major implications for public health. In view of

this, we consider now is the appropriate time that we should together formally retract the interpretation placed upon these findings in the paper." John O'Leary, the Irish pathologist who had
found measles virus in the intestines of autistic children, said he
was "shocked and disappointed" by Wakefield's behavior.

The British media also felt cheated. Prior to Deer's revela-
tions, Wakefield had been a hero, willing to take on powerful,
self-serving forces such as public health officials, drug companies, and an entrenched medical establishment. Now some in
the British press saw him as a shill for plaintiffs in a massive
lawsuit. Headlines in *The Guardian* and *Sunday Times* read,
"MMR Scare Doctor Got Legal Aid Fortune," "MMR Scandal
Doctor May Face Professional Misconduct Charges," and "Is
This Doctor a Hero or a Health Risk?"

· · · ·

Although the British press saw Brian Deer's allegations
as tainting the notion that MMR caused autism, scientific studies
had already gone a long way toward refuting Wakefield's claims.
In his original *Lancet* paper Wakefield admitted that he had not
proven an association between MMR and autism. To determine
whether MMR caused autism, researchers would have to perform
a series of epidemiological studies, comparing hundreds of thousands of children who did or did not receive the vaccine. Brent
Taylor's study, published in the *Lancet* and later presented to Dan
Burton's Committee on Government Reform, was the first to
clearly show that vaccinated children were not at increased risk.
Others followed. Scientists from Helsinki University in Finland
examined the records of 2 million children and found no evidence
that MMR caused autism. Researchers from Boston University
School of Medicine examined the medical records of more than 3
million children; although they found an increase in the rates of
autism, that increase wasn't associated with the children's getting
MMR. Loring Dales and Natalie Smith, from the Department of
Health Services in Berkeley, California, also found that increasing
rates of autism occurred independent of the use of the MMR vac-

cine. Brent Taylor extended the study he had presented to Dan
Burton by following children for longer periods of time, finding
that the MMR vaccine didn't cause autism up to five years after it
was given. Frank DeStefano, from the Centers for Disease Control and Prevention (CDC), performed a study in metropolitan
Atlanta showing that children with autism were not more likely
to have received MMR. Scientists from Denmark examined the
medical records of more than 500,000 children; some had gotten
the MMR vaccine and some hadn't. Again the risk of autism was
the same in both groups. The American Academy of Pediatrics
(AAP) and the Institute of Medicine (IOM; an independent research group within the National Academy of Sciences) reviewed
these and other studies and reached the same conclusion: MMR
didn't cause autism. Finally, public health officials in Japan, who
had discontinued MMR following Wakefield's report, didn't see
any decline in the rates of autism; instead, the number of children
with the disorder continued to soar.

More important, Wakefield's explanation of how MMR caused
autism didn't hold up. Wakefield had claimed the measles vaccine virus in MMR had caused a chronic infection of the intestines, allowing harmful proteins to enter the bloodstream and
damage the brain. However, other researchers couldn't find any
evidence that measles vaccine caused a chronic infection in autistic children. Like O'Leary, these investigators used PCR to
detect measles virus genes. One of the Canadian researchers, Brian
Ward, said, "In our hands, the [data] published by [O'Leary]
yielded many PCR 'positive' results that turned out to be falsely
positive on closer examination. These data are a direct refutation of the reports of persistence of measles virus in the tissues
of autistic children. We are hopeful that this paper will simply
put a quiet end to the debate surrounding this topic." Later, independent examiners tested the competence of O'Leary's laboratory (Unigenetics) by asking it to identify which coded samples
contained measles virus. Unigenetics failed the test. One of the
examiners said the "record of performance [of O'Leary's testing
laboratory] would not be acceptable for certifying a clinical

laboratory." Stanley Plotkin, an emeritus professor at the University of Pennsylvania School of Medicine and the inventor of the rubella vaccine, concluded, "O'Leary's data provided Wakefield with the linchpin evidence that was central to his theory. If that falls away, all you have is a hypothesis unsubstantiated by anything. The whole thing falls into the water." In 2003, John O'Leary admitted his PCR evidence "in no way established any link between the MMR vaccine and autism."

Wakefield's contention that MMR caused a leaky gut also suffered several blows. First, Wakefield believed it was the combination of measles, mumps, and rubella vaccine that was causing intestinal disease. He didn't think giving the vaccines separately would be a problem. But studies later showed that children given MMR were *not* more likely to develop bowel problems than children given the vaccines separately.

Second, a closer look at Wakefield's original *Lancet* study showed that only one in five children had bowel problems *before* they developed autism. If a leaky gut were the cause, bowel symptoms should have preceded, not followed, the first symptoms of autism.

Third, Wakefield believed harmful proteins were entering a damaged gut. But if the gut were damaged, the leak should go both ways, from the gut to the bloodstream and from the bloodstream to the gut. If autistic children had leaky guts, then their stools should have contained large amounts of proteins normally found in the blood. But this wasn't the case. Michael Gershon, professor and chairman of the Department of Anatomy and Cell Biology at Columbia University, commented on this paradox: "If the presumed 'leak' were large enough to permit [proteins] to enter the body in significant amounts, then body proteins would be expected to move simultaneously in the opposite direction into the [center] of the bowel."

Fourth, Wakefield believed that the combination of measles, mumps, and rubella vaccines had simply overwhelmed a young child's developing immune system. But the challenge of these three viruses is minuscule compared with that of vaccines that had been developed and administered in the past. Viruses are made of proteins; larger viruses contain more proteins than smaller viruses. Measles, mumps, and rubella viruses are quite small, containing ten, nine, and five proteins, respectively. Smallpox, on the other hand, is one of humankind's largest viruses; it contains about 200 proteins. So, although smallpox is only one vaccine and the combination MMR vaccine contains three, the number of immunologic challenges from the smallpox vaccine is much greater than from MMR (200 versus 24). Indeed, children confronted more immunologic challenges by receiving only the smallpox vaccine 100 years ago than they do while receiving 14 different vaccines today (200 versus 153). If immunological overload were the cause of autism, with fewer immunologic challenges in modern vaccines, rates of autism should be decreasing, not increasing.

Finally, the work of Vijendra Singh came under closer scrutiny. Singh had testified in front of Dan Burton's committee that MMR had induced unusual levels of antibodies against measles virus as well as antibodies against nerve cells. But a closer look at Singh's science revealed two critical flaws: children with autism didn't have evidence of nerve cell damage and, according to measles experts, the test that Singh had used to detect measles antibodies didn't detect them.

· · · ·

ON FEBRUARY 18, 2004, ANDREW WAKEFIELD SAT IN RICHARD Horton's office after hearing Brian Deer's allegations against him. During the six years that had passed since he had published his paper in the *Lancet*, scores of scientific studies had failed to support his contention that MMR caused autism. Horton knew the game was coming to an end; he remembered the moment: "I turned to Wakefield and remarked, more as a way to break the silence than as a comment in need of a reply, 'It seems like this whole affair is coming to a head.' It was a baleful understatement, one that failed to match the occasion—the imminent implosion of work that had divided parents from doctors, and doctors from their colleagues. Perfectly aware of these realities, Wakefield looked past me, expressionless."

For Andrew Wakefield, the news would only get worse.

· · · ·

IN THE UNITED STATES, IF PEOPLE BELIEVE THEY ARE HARMED BY a medical product, personal-injury lawyers represent them. The same is true in England, with one important difference. When lawyers in the United States gather evidence to support a claim, they pay for it themselves. A typical example is described in Jonathan Harr's *A Civil Action*. Parents in Woburn, Massachusetts, believed their children got cancer from drinking water contaminated by chemicals from a nearby tanning factory, W. R. Grace. Convinced that Grace had polluted the water supply, they asked a lawyer named Jan Schlichtman to represent them. (Schlictman was played by John Travolta in the movie version.) Schlictman's law firm spent hundreds of thousands of dollars gathering medical evidence and interviewing toxicologists to support the parents' claim. In contrast, the United Kingdom has a program called the Legal Services Commission that provides money to lawyers to investigate claims. The commission, which has a fund of more than $1 billion, employs a panel of advisers to determine which claims are worth pursuing. When parents in the United Kingdom believed their children had been harmed by MMR, they sought out Richard Barr, who then applied for and won money from the commission.

In October 2006, more than two years after he had claimed that Andrew Wakefield had received $100,000 from Richard Barr, Brian Deer asked the Legal Services Commission for information under the Freedom of Information Act. He wanted to know how Barr had spent the money. On December 22, 2006, Deer received a report titled "Freedom of Information Act Request: MMR Multi-Party Action." After he read the report, Deer realized he had grossly underestimated the amount of money that had been spent to support the case against MMR. The commission had provided $30 million: $20 million went to Richard Barr's law firm; $10 million went to doctors and scientists.

To determine which investigators to support, Barr's law firm set up a sixteen-person "Science and Medical Investigation Team." The team's leader had graduated from college with a Bachelor of Science degree and, according to the pamphlet used to recruit parents for the class-action lawsuit, had "an encyclopedic knowledge of medical matters." Two members of the team had worked as laboratory technicians, one with Wakefield. Two others were nurses; the rest were lawyers. Not one member of Barr's team had ever led a scientific investigation or completed graduate-level training in immunology, virology, molecular biology, microbiology, statistics, or epidemiology.

· · · ·

BARR'S SCIENCE AND MEDICAL INVESTIGATION TEAM GAVE $800,000 to Andrew Wakefield to support his research. When Wakefield had been confronted by journalists two years earlier, following Deer's claim that he had received $100,000, he argued that the sum was closer to $50,000. Now, with a document from the Legal Services Commission in front of him, Wakefield didn't deny that the amount was far greater. "This work involved nights, weekends, and much of my holidays," he explained, "such that I saw little of my family during this time. I believe and still believe in the just cause of the matter under investigation."

Wakefield had other financial interests. Several months before his press conference announcing the dangers of MMR, he had filed a patent application for a safer measles vaccine. In the application, Wakefield had stated, "I have now discovered a combined vaccine–therapeutic agent which is not only most probably safer to administer to children and others by way of vaccination, but which also can be used to treat [autism] whether as a complete cure or to alleviate symptoms." The coholder of Wakefield's patent was the Royal Free Hospital, which had prepared a videotape for the press conference claiming that MMR vaccine was unsafe. Wakefield's patent contended that measles virus, white blood cells from mice, and pregnant goats could be used to make a safer measles vaccine. Tom MacDonald, a professor at Southampton University and Britain's foremost gut immunologist, called the concoction "total bollocks." Wakefield also started two companies, Immunospecifics Limited and Carmel Healthcare (named after his wife), to sell diagnostic kits to parents of autistic children. But his products never progressed beyond the concept stage.

Wakefield wasn't the only investigator to benefit from Barr's largesse. Barr's team had given John O'Leary more than $1 million to perform his studies. But the money hadn't gone directly to O'Leary. Rather, Barr had given it to Unigenetics Limited, a company that O'Leary had set up on the campus of Coombe Women's Hospital to test the intestinal samples provided by Wakefield. When the money that Unigenetics had received was made public, O'Leary said, "I cannot confirm the fees paid for this testing, but will consult with our accountants and endeavor to do so. I should also emphasize that I personally have not been paid anything by the Legal Services Commission, although I did receive fees from Unigenetics Limited." John O'Leary was the director and the major shareholder of Unigenetics.

Kenneth Aitken, the clinical pathologist who had publicly supported Wakefield, had received $400,000. Before receiving the money, Aitken had edited *Children with Autism: Diagnosis and Intervention to Meet Their Needs*, an important and well-respected textbook in the field. In the book, Aitken had postulated the cause of autism: "It now seems certain that the brains of persons who become autistic in their early childhood already had microscopic faults in their development in early intra-uterine life, probably first expressed among cells of the early embryo, in the first month." But after receiving money from Barr, Aitken appeared to change his opinion, now believing that autism might be caused by MMR. One month after Wakefield's publication in the *Lancet*, Britain's Medical Research Council appointed Aitken to a thirty-seven-member board to determine whether Wakefield's science justified a change in vaccination policy. Aitken was the only member of the council to vote that it did (and the only member to have received money from Barr). Following an unrelated scandal, Ken Aitken resigned from his position in Edinburgh.

Walter Spitzer, the Canadian epidemiologist who had testified before Dan Burton's committee that the odds favored a link between MMR and autism, had received $30,000. Spitzer's assessment of Wakefield's claims didn't go unnoticed by his Canadian colleagues. Angered by his claims, Noni MacDonald, the dean of medicine and professor of pediatrics and medicine at Dalhousie University in Halifax, Nova Scotia, said, "I am embarrassed that he is an emeritus professor at McGill. I think he better go back and look at proper causality assessment before he makes that kind of statement. I would flunk him."

Arthur Krigsman, the gastroenterologist from the New York University School of Medicine, had declared during Burton's hearings that his findings were "independent" from Wakefield's; he had received $30,000.

Marcel Kinsbourne and John Menkes, the California neurologists who had been the first to support Wakefield's hypothesis in the American press, had received $800,000 and $90,000, respectively. Toward the end of their careers, both men had become expert witnesses for lawyers suing vaccine makers. (Marcel Kinsbourne would be heard from again.)

John March, the veterinarian who had claimed that animal vaccines were tested more extensively than MMR, had received $180,000. As the parent of an autistic child, March had been drawn into the controversy by Wakefield's claims that MMR hadn't been adequately tested before licensure. Michael Fitzpatrick, the author of *MMR and Autism* and father of an autistic child, later talked to John March about his odyssey. "[March] tells an interesting story of how he presented his data, carried out on behalf of the litigation, to a weekend meeting presided over by Richard Barr and Wakefield in a plush country hotel," recalled Fitzpatrick. "When he told them that there was no difference between the children with autism and controls, he suddenly found that the meeting had moved on to a different subject. It was a Damascene conversion for him. He realized that Wakefield could not hear negative results." March also talked to Fitzpatrick about the huge sum of money he had received from Richard Barr's team. "I was interested when it emerged how much he had earned from the litigation," recalled Fitzpatrick, "because he told me that he had been doing the lab work gratis at first, until he discovered that everybody else was claiming vast expenses and was advised that he should do likewise." "There was a huge conflict of interest," said March. "It bothered me quite a bit because I thought, well, if I'm getting paid for doing this, then surely it's in my interest to keep going as long as possible. The ironic thing is [Barr's team] was always going on about how, you know, how we've hardly got any money compared with the other side who are funded by large pharmaceutical companies. And I'm thinking, judging by the amounts of money you're paying out, the other side must be living like millionaires."

Evan Harris, a Liberal-Democrat Member of Parliament, later commented on the money paid by the Legal Services Commission to investigate MMR. "Those figures are astounding," he said. "The lawsuit was an industry, and an industry peddling what turned out to be a myth."

Richard Barr defended the notion that a team of lawyers could lead a scientific investigation. "We became versed in the key scientific literature, filling yards of shelf space in the offices with prints of relevant papers," he wrote. "Our office at one stage was full of cardboard urine containers and storage bottles. And we had adventures too." One of Barr's adventures involved flying autistic children across the Atlantic Ocean. Because British hospitals often refused to perform spinal taps on autistic children, Barr flew seven autistic children from England to Detroit, where spinal taps were performed in a private clinic. But despite collecting gallons of blood, urine, and spinal fluid from hundreds of autistic children, and despite receiving $10 million, the researchers and clinicians supported by Barr's team never found the chronic measles infections, leaky guts, or brain-damaging proteins Andrew Wakefield had imagined.

• • • •

IN SEPTEMBER 2003—AFTER RICHARD BARR ASKED FOR ANOTHER $20 million to pursue his case against vaccine makers—the Legal Services Commission withdrew its support. The commission, convinced that medical research had failed to provide a link between MMR and autism, said it would "not be correct" to spend more money on a trial that "is very unlikely to succeed." Claire Dodgson, the chief executive of the commission, said, "I appreciate that this decision will come as a great disappointment to the parents involved. I sympathize with their situation. Their children are clearly ill and they genuinely believe the MMR vaccine caused their illness. However, this litigation is very unlikely to prove their suspicions." Barr was crushed. He appealed the commission's decision to the Funding Review Committee, but was turned back. On February 27, 2004, he tried one more time, appealing to the federal Judicial Review Committee, again without success. Realizing that further appeals would be unlikely to succeed, Richard Barr dropped his case against pharmaceutical companies, a case that had been expected to go to court in April 2004.

Barr later commented on the decision of the Legal Services Commission to abandon his clients. "I have some idea of what the anonymous sculptor of the Venus de Milo must have felt," he said. "He had chipped away for years, slowly transforming a piece of the best available marble into a work of unimaginable beauty. Then, just when it was almost ready to be handed over, a well meaning cat rubbed against the base, the whole thing toppled over, and both arms fell off. I don't know what language the sculptor spoke, but if there was a word equivalent to 'bugger' or worse than that I'm sure he uttered it. And [that's] what I said and thought when the news came through to me on a slow train journey back to Norfolk [England], that the Legal Services Commission had, even after appeal, pulled funding from the

MMR cases." Unlike the 1,300 families he represented, Richard Barr had been compensated for his efforts, his law firm having received more than $20 million.

The case against MMR was the first in England's history in which the Legal Services Commission financed scientific research. And it will probably be the last. The commission concluded: "In retrospect, it was not effective or appropriate for [us] to fund research. The courts are not the place to prove medical truths." The commission reasoned that science directed by a team of personal-injury lawyers wasn't likely to be the best kind of science.

• • • •

IN DECEMBER 2001, THE ROYAL FREE HOSPITAL ASKED ANDREW Wakefield to resign. "I have been asked to go because my research results are unpopular," he said. "I did not wish to leave but I have agreed to stand down in the hope that my going will take the political pressure off my colleagues. They have not [fired] me. They cannot. I have not done anything wrong. Losing a London hospital teaching job doesn't do much for my [resumé] but there are bigger issues at stake. What matters now most is what happens to these children."

No longer part of a major university or teaching hospital, Andrew Wakefield took his case directly to the press and public, disregarding scientists who disagreed with him. In February 2004, soon after leaving England, he became the director of research at the International Child Development Resource Center in Melbourne, Florida. Two physicians, Jeff Bradstreet and Jerold Kartzinel, direct the center and call it the home of the "Good News Doctor." In his biography, Kartzinel claims that his youngest son's autism developed immediately after receiving MMR. One year later, Wakefield moved to Austin, Texas, to work at the Thoughtful House Center for Children, which advertises, "If your child is diagnosed with autism, the next thing to know is that autism is treatable."

In September 2005, the General Medical Council (GMC) in London, the agency responsible for licensing and monitoring physicians, charged Andrew Wakefield with several counts of professional misconduct. It claimed Wakefield had called for a boycott of MMR without clear evidence that the vaccine caused harm; recruited children to his study through anti-vaccine pressure groups; retained and used human specimens without consent; failed to answer questions from the government's chief medical officer about the source of his funding; subjected children to "unnecessary and invasive investigations"; and gave five pounds (ten dollars) directly to children attending his son's birthday party to collect their blood. The consequences of this two-year investigation could be that Andrew Wakefield is permanently barred from practicing medicine in England. Wakefield has contested these charges, denying any wrongdoing.

· · · ·

During his 60 *Minutes* interview with Ed Bradley in November 2000, Wakefield had said, "I would have enormous regrets if [my theories] were wrong and there were complications or fatalities from measles." Wakefield was right in acknowledg-

Andrew Wakefield, pictured with his wife Carmel, is flanked by supporters on his way to a hearing before Britain's General Medical Council on charges of misconduct, July 16, 2007 (courtesy of Getty Images).

ing that some parents might soon watch children suffer and die from measles, but he had overestimated his capacity for regret. Although study after study showed MMR didn't cause autism, Wakefield remains unrepentant, wedded to a belief he considers irrefutable. Seeing himself as a champion of children, he asks: "Should we stop, should we go away, should we stop publishing because it is inconvenient? I've lost my job. I will never practice medicine in [England] again. There is no up side to this. But if you come to me and say, 'This has happened to my child,' what's my job? What did I sign up to do when I went into medicine? I'm here to address the concerns of the patient. There's a high price to pay for that. But I'm prepared to pay it." Since 2005, when Andrew Wakefield first traveled to Thoughtful House in Austin, he has treated more than 2,000 autistic children, still performing colonoscopies when he believes they are necessary.

Michael Fitzpatrick never saw Wakefield as a humble servant of the people. As the father of an autistic son, Fitzpatrick is angry that Wakefield has diverted attention and research away from the real cause or causes of autism, adding his own postscript to the Wakefield saga: "Although Dr. Wakefield's self-consciously humble posture is a popular departure from the traditional image of the paternalistic doctor, it raises some questions. If it is true that he has learned everything he knows about autism from parents, this suggests that his knowledge is very limited; parents are in no position to acquire the broad scientific understanding of autism required by a medical researcher. If it is not true, which is more likely, it is merely insincere. Furthermore, Dr. Wakefield appears to be highly selective in his listening: while he hears parents who endorse his views, he remains deaf to parents who do not. Nor does he appear to listen to the vast majority of his scientific and medical peers. For all Dr. Wakefield's talk about humility, his continuing public promotion of the anti-MMR cause in face of the weight of evidence to the contrary does not suggest a surfeit of this virtue."

. . . .

FOLLOWING ANDREW WAKEFIELD'S PUBLICATION IN THE *LANCET*, rates of immunization with MMR declined and measles outbreaks swept across the United Kingdom and Ireland; hundreds of children were hospitalized and four children died from a disease that could easily have been prevented. Further, the campaign against MMR caused, according to Michael Fitzpatrick, "an enormous wave of unnecessary anxiety among parents facing decisions about new immunizations, slowed the introduction of new childhood vaccines in Britain, encouraged the emergence of sleazy clinics selling single-agent vaccines, and added an additional burden of guilt and self blame to parents." Unfounded fears of the MMR vaccine clearly damaged the public's health. Who, if anyone, should be held accountable?

Andrew Wakefield, the Royal Free Hospital, the British media, the Legal Services Commission, and the *Lancet*'s editor-in-chief, Richard Horton, all played a role in the events that followed.

Wakefield published his *Lancet* paper based on the findings of eight children with autism. He knew that the only way to prove his contention was to show that autism was more common in children who had received MMR. And he knew that he hadn't done that study. Correctly, Wakefield explained the limits of his study in the discussion section of his paper: "We did not prove an association between measles, mumps, and rubella vaccine and [autism]." It would have been more accurate if he had said he hadn't provided *any* evidence that MMR caused autism and had merely reported the convictions of the parents of eight autistic children. But Wakefield came to his press conference with his own public relations firm in tow and, throwing caution to the wind, told the press and the public that he had likely found a cause of autism. When study after study refuted his theory, he remained unrepentant. "I don't think he's an innocent in this," said David Salisbury, the director of immunization for the Department of Health in the United Kingdom. "He knew exactly what he was doing. And throughout he has never shown the slightest contrition for what he has caused.

He's had more than enough opportunities to say to the world, 'I deeply regret the fact that I, acting out of the best of interests, got this wrong and now realize the consequences of what has happened.' He's never done this."

During Wakefield's press conference, the Royal Free Hospital showed a videotape warning of the possible dangers of MMR. The hospital was also a coholder on Wakefield's patent application for a safer measles vaccine. And members of the Royal Free, including John Walker-Smith and Arie Zuckerman, willingly participated. "They could have [and] should have restrained Wakefield at an earlier stage, or at least have stopped the press conference grandstanding," recalled Michael Fitzpatrick. "But it seems they were all in some way captivated by him, by his offer to end their careers in a blaze of glory. It almost seems that they looked to Wakefield as the savior of the Royal Free, a folly that ended up damaging everybody. Like many aspects of Wakefield's story, it has elements of a Greek tragedy."

Following Wakefield's *Lancet* publication, the media trumpeted his claims. Journalists like Melanie Phillips of the *Daily Mail*, Lorraine Fraser of the *Sunday Telegraph*, and Heather Mills of *Private Eye*, wrote of unsavory drug company influence and health officials asleep at the switch. They painted Andrew Wakefield as a hero, a man of the people, willing to confront forces bent on crushing him. It was great theater. The deterioration of trust in vaccines could never have happened without them. "The media coverage told parents not only what to think, but also how to think about the MMR vaccine," wrote Tammy Boyce, author of *Health, Risk and News: The MMR Vaccine and the Media*. Boyce argued that the media's ritualistic mantra of balance—equally weighing one man's speculations with studies that clearly exonerated MMR—created a "charade of objectivity."

The Legal Services Commission gave $30 million to Richard Barr to investigate the case against MMR. The money spent by Barr's team funded bad science that was never reproduced by other scientists. When these poorly designed and poorly executed studies were exposed, the commission's board members questioned whether it was appropriate to allow personal-injury lawyers to direct scientific research. But contrition came far too late, the damage done.

Richard Horton, the editor-in-chief and gatekeeper of the *Lancet*, was probably in the best position to prevent the harm caused by Andrew Wakefield's paper. With a database of 8,000 expert reviewers, a circulation of 40,000, and more than 1 million online users, the *Lancet* is the longest-running general medical journal in the world, and one of the most influential. Launched in 1823 by Thomas Wakley, a doctor, coroner, and Member of Parliament, the *Lancet* was the first journal to describe the value of blood transfusions, antiseptics, and penicillin. The *Lancet* was also known for its muckraking. Wakley, who named the journal for the sharp instrument used to puncture festering boils, used his editorial position to rail against what he saw as corruption, nepotism, and incompetence in the medical establishment. Richard Horton has followed Thomas Wakley's lead. Journalist Nicole Martin noted that Horton's "firebrand style and penchant for challenging vested interests in medicine met with admiration and criticism among medical editors." Another journalist later remarked, "Secretly, I admire him. But I do wonder if he is slipping sideways into journalism rather than scientific editorship."

Because Wakefield's paper was inconclusive, Horton simultaneously published an editorial from Robert Chen and Frank DeStefano, vaccine safety experts from the CDC in Atlanta, warning of the study's limitations. By including Chen and DeStefano's editorial, Horton believed he had bracketed Wakefield's paper with the necessary caveats. But doctors, scientists, immunologists, virologists, and public health officials, angry that the journal had published such poor science, immediately flooded the *Lancet* with letters. Horton defended his decision: "There was no question in my mind that, subject to external peer review and editorial debate, we should publish this work. The description of what seems to be a new syndrome and its relation to possible environmental triggers was original and would certainly interest our readers. Recent history tells us that full disclosure of new data is preferable to well-meaning censorship." Critics countered that the *Lancet* published only 5 percent of the 10,000 papers it received every year, wondering whether rejection of the other 95 percent was an act of censorship or editorial judgment. Horton also hadn't anticipated that Wakefield would hold a press conference warning parents against MMR, even though Wakefield had done exactly that several years earlier when he announced that measles vaccine caused Crohn's disease, a claim he later retracted in the face of overwhelming evidence to the contrary.

Richard Horton later published two books discussing his role in the controversy, *MMR Science and Fiction: Exploring the Vaccine Crisis* and *Second Opinion: Doctors, Diseases, and Decisions in Modern Medicine*. Five years after he had published Wakefield's paper, Horton was unrepentant. "There [is] an unpleasant whiff of arrogance in this whole debate," he said. "Can the public not be trusted with a controversial hypothesis? The view that the public cannot interpret uncertainty indicates an old-fashioned paternalism at work. The public is entitled to know as much as possible." But by ignoring the criticisms of several reviewers, the warnings of an accompanying editorial, Wakefield's history of holding press conferences, a British press primed for controversy, and a public distrustful of public health officials, Richard Horton allowed parents to question the safety of a vaccine based on flimsy, irreproducible data. The loss of public trust that followed was entirely predictable. "It was a stunning error of judgment," opined David Salisbury. "It is hard to believe that the paper was properly reviewed. On the link with MMR, it was a complete mess, and had a chance of being correct that was about zero. [Horton] bears a considerable burden of responsibility."

Learning little from his encounter with Andrew Wakefield, Richard Horton has published papers in the *Lancet* claiming that genetically modified foods damaged rat intestines, silicone breast implants induced harmful antibodies, and casualties sustained during the U.S.-led invasion of Iraq totaled 655,000 (about ten times the actual number). Like Wakefield's paper, all of these assertions garnered enormous media attention for his journal, and all have been clearly refuted.

Despite massive educational campaigns by public health officials, measles immunization rates in the United Kingdom never fully recovered. In 2006, more cases of measles were reported than in any single year since 1995. Between June and August 2007, the number of measles cases tripled.

• • • •

DESPITE THEIR PROFESSED GOOD INTENTIONS, DOUGLAS BIKLEN (facilitated communication), Karoly Horvath (secretin), and Andrew Wakefield (MMR vaccine) had proven to be false prophets in the quest to find a cause and a cure for autism. In the next few years, parents would turn their attention to another vaccine component and yet another group of unlikely heroes.

CHAPTER 4

A Precautionary Tale

The scars of others should teach us caution.

—St. Jerome

Many parents had been persuaded by the MMR controversy that vaccines caused autism. When studies exonerated MMR, they reasoned it must be something else in vaccines causing the problem. It wouldn't be long before they believed they had found it.

• • • •

Frank Pallone is a congressman who represents New Jersey's Sixth District. First elected in 1988, he has been a passionate supporter of Native Americans, working to protect the sovereignty of tribal governments. He's also an environmentalist. Because Pallone lives in a district that includes a string of towns on the Jersey shore, he's particularly worried about contamination of fish with mercury. In 1997, Pallone attached a simple amendment to an FDA reauthorization bill. Only 130 words long, the amendment would soon lead to chaos. Pallone gave the FDA two years to "compile a list of drugs and foods that contain intentionally introduced mercury compounds and [to] provide a quantitative and qualitative analysis of the mercury compounds in the list." The bill—the FDA Modernization Act of 1997—was signed into law on November 21, 1997. Few in the press or the public took notice.

One year passed. On December 14, 1998—eleven months before the congressional deadline—the FDA put a notice in the *Federal Register* again asking food and drug makers to list the amounts of mercury in their products. The *Federal Register*—a daily publication of federal regulations, legal notices, presidential proclamations, executive orders, funding priorities, grant deadlines, and meetings—didn't gain the attention of pharmaceutical companies. No one complied. So on April 29, 1999, the FDA made yet another, more specific announcement. A few weeks later, FDA officials received the information they needed. What they found concerned them. By six months of age, infants could receive as much as 75 micrograms of mercury from three doses of the diphtheria-pertussis-tetanus (DPT) vaccine, 75 micrograms from three doses of the *Haemophilus influenzae* type b (Hib) vaccine, and 37.5 micrograms from three doses of the hepatitis B vaccine—a total of 187.5 micrograms of mercury. (A gram is approximately the weight of one-fifth of a teaspoon of salt. A microgram is one-millionth of a gram.) Everyone at the FDA knew that environmental mercury (known as methylmercury) could cause serious, permanent, and occasionally fatal damage to the nervous system—a lesson learned from an event that had occurred decades earlier at Minamata Bay.

· · · ·

DURING WORLD WAR II, JAPAN'S MOST SOUTHERN ISLAND, KYUSHU, was home to a chemical company that dumped its waste into Minamata Bay. Within months, fish began to float in the bay. Then cats started jumping into the sea and dying—local residents called them "cat suicides." Seagulls fell from the sky. By the early 1950s, people began to suffer unexplained numbness in their hands and feet, muscle weakness, and a narrowing of their field of vision; some had difficulty walking, swallowing, and hearing; others trembled uncontrollably. In extreme cases, insanity, paralysis, coma, and death followed. By 1956, epidemiologists had found the cause: mercury dumped into the water had been concentrated in shellfish, poisoning consumers. Kyushu's chemical factory continued to pour mercury into industrial wastewater until 1968. By 2001, mercury had poisoned 3,000 Japanese citizens and killed 600.

．．．．

FDA OFFICIALS KNEW THAT MERCURY COULD DAMAGE THE NERvous system. When they first responded to Frank Pallone's directive, they also knew that mercury-based preservatives had been used in vaccines for decades. They were the ones who had put it there.

Between 1900 and 1930, companies packaged vaccines almost exclusively in multidose vials, with a typical vial containing ten doses. Because a large percentage of the cost of vaccines is determined by its packaging—sterile glass vials, rubber stoppers, metal tops, and labels—as well as the labor required to fill each vial, multidose vials allowed vaccines to be made much less expensively. Doctors kept these vials in refrigerators in their offices, often for months at a time. To give a vaccine, they would insert a needle through the rubber stopper, pull the liquid up into a syringe, and inject it. Unfortunately, by constantly violating the rubber stopper with a needle, doctors and nurses occasionally contaminated the vial with bacteria. In 1916, in Columbia, South Carolina, a batch of typhoid vaccine contaminated with bacteria caused seventy severe reactions and four deaths. And in Queensland, Australia, in 1928, twelve children injected with a contaminated diphtheria vaccine died from bacterial abscesses and bloodstream infections. By the 1940s, most multidose vials of vaccines contained preservatives to prevent contamination.

The choice to use mercury as a preservative was based on what was available. Before antibiotics were discovered, doctors relied on antiseptics to kill bacteria, the most popular of which was Joseph Lister's carbolic acid. Other scientists championed other antiseptics. Robert Koch, the father of the germ theory of disease, preferred mercury chloride, which was unfortunately an irritant. Early in the twentieth century, a new, more effective, less toxic derivative of mercury came into favor: ethylmercury. (The most popular mercury-containing antiseptic for home use was called Mercurochrome, a bright- orange-colored liquid used on scrapes and cuts.) Synthesized by the pharmaceutical company Eli Lilly, ethylmercury soon became the most commonly used preservative in vaccines. But rather than use ethylmercury in its pure form, Lilly combined it with thiosalicylate and called it thimerosal.

Before they put thimerosal in vaccines, Lilly scientists first studied it. They found that thimerosal was much better at killing bacteria than other antiseptics. They also found that they could inject hundreds of milligrams of it into rabbits and rats without any harmful effect. Then they gave it to people. In 1929, during an outbreak of bacterial meningitis in Indiana, Lilly scientists gave thimerosal to doctors to treat the infection. It didn't work. (It would be another six years before the first antibiotic, sulfa, entered the United States.) Although thimerosal didn't treat meningitis, doctors found that it was safe. Adults injected with 2 million micrograms of thimerosal didn't suffer symptoms of mercury poisoning; the amount was 10,000 times greater than the FDA later found babies had received in vaccines.

After scientists added thimerosal to vaccines, infections caused by contamination of multidose vials virtually disappeared. For decades, the preservative was used without a second thought. But as health officials added more vaccines to the routine schedule, children received more and more mercury.

• • • •

ONCE FDA OFFICIALS FOUND EXACTLY HOW MUCH MERCURY WAS in vaccines, they had to determine whether it was a safe amount. So they consulted safety guidelines from four sources: their own agency, the Environmental Protection Agency (EPA), the Agency for Toxic Substances Disease Registry, and the World Health Organization (WHO). What they found surprised them: safety guidelines didn't exist. Guidelines were in place for environmental mercury (methylmercury). But vaccines contain ethylmercury. (Scientists preferred ethylmercury over methylmercury because it is eliminated from the body much faster.) Although the distinction between methylmercury and ethylmercury—a difference of only one carbon molecule—sounds trivial, it's not. An analogy can be made between ethylalcohol, contained in wine and beer, and methylalcohol, contained in wood alcohol. Wine and beer can cause headaches and hangovers; wood alcohol causes blindness.

In mid-June 1999, two months after they had determined how much mercury was in vaccines, FDA officials held a meeting to discuss what they had found. Among others, the FDA invited Neal Halsey, a fifty-seven-year-old pediatrician from Johns Hopkins Medical School, to attend. Halsey had been a vaccine insider for more than two decades, serving on advisory committees to the CDC and the AAP. Because these two organizations are almost solely responsible for recommending vaccines in the United States, few could claim greater responsibility for deciding which vaccines American children got and when they got them than Neal Halsey. For several hours Halsey listened while FDA officials pored over reams of data showing that the amount of *ethyl*mercury children received from vaccines had exceeded that recommended by the EPA for *methyl*mercury. "My first reaction was simply disbelief," said Halsey, "[as] was the reaction of almost everybody involved in vaccines." Halsey felt misled by how mercury had been listed in package inserts. "In most vaccine containers," said Halsey, "thimerosal is listed as a mercury derivative, a hundredth of a percent. And what I believed, and what everybody else believed, was that it was truly a trace, a biologically insignificant amount. My honest belief is that if the labels had had the mercury content in micrograms, this would have been uncovered years ago. But the fact is no one did the calculations." Halsey wondered whether pharmaceutical companies, in an effort to cover up the amount of mercury in vaccines, had purposefully made it difficult for federal advisers to know exactly how much mercury children were getting. "Gradually it came home to me that maybe there was some real risk to children," he said. Feeling deceived, Halsey was angry and frustrated.

Halsey had also been influenced by an event that had occurred earlier that summer. While canoeing with his family in Maine, he had come across a sign that advised fishermen to "protect your children—release your catch." Park officials were worried that fish contained quantities of mercury that might harm children. If the government was warning people about eating fish with mercury, Halsey wondered, "Does it make sense to allow it to be injected into infants?"

. . . .

ALTHOUGH HALSEY WAS CONCERNED ABOUT THE AMOUNT OF mercury in vaccines, several studies available at the time were reassuring. First, studies in the Seychelles, a group of islands in the Indian Ocean, had shown that infants and young children who consumed large amounts of mercury in fish didn't develop neurological problems. Second, levels of mercury contained in vaccines did not exceed safety guidelines recommended for methylmercury by the FDA, the agency charged with determining the safety of vaccines. Third, EPA guidelines were based on methylmercury, which is eliminated from the body far more slowly than ethylmercury and is, as a consequence, much more likely to accumulate and cause harm. Fourth, EPA safety levels were determined by mercury exposures to the fetus, whose brain and other organs are still developing. By using levels considered to be harmful to the fetus to determine those likely to be harmful to infants, and by extrapolating guidelines for methylmercury to ethylmercury, the EPA had overstated the risk. Finally, several European countries had already removed thimerosal from vaccines without an obvious drop in the rates of neurological problems; but the effect of this change had never been formally studied.

Despite the reassuring information, Halsey wasn't reassured. That's because he couldn't get the results of one particular study out of his mind. Between 1986 and 1987 investigators had studied children who lived on the Faroe Islands, a small group of islands in the North Sea. Children of the Faroes ate whale meat that was heavily contaminated with methylmercury. Researchers examined 1,000 Faroes children, determining the amount of mercury in their umbilical cords and hair. Seven years later they administered a series of tests. They found that children exposed to fairly low levels of mercury while in their mother's womb didn't score as well on tests of attention, memory, language, and (to a lesser extent) movement and sight. Unlike studies in the Seychelles, this research showed that relatively small quantities of mercury in food could affect children. (Later, researchers realized that fish from the Faroe Islands were also contaminated with polychlorinated biphenyls [PCBs], confounding the interpretation of their study.) Halsey read this study over and over again. Although he didn't think that mercury caused autism, he gradually became convinced that it might be causing subtle neurological or psychological problems.

Halsey knew that the best way to determine whether mercury in vaccines was harmful was to compare children who had received thimerosal-containing vaccines with those who hadn't. Although these studies would come soon, at the time of the FDA meeting they didn't exist. So Halsey decided to evoke the precautionary principle, exercising caution in the absence of evidence. Unlike Andrew Wakefield, who had made precipitous recommendations against the MMR vaccine, Halsey would not recommend the removal of a component of vaccines necessary to protect children from potentially fatal infectious diseases. He would merely propose that vaccine makers switch to single-dose vials free of preservatives. Although this recommendation would make vaccines more expensive, it was to Halsey a price worth paying given the potential consequences.

. . . .

HALSEY HAD ONE MORE WEEK AS HEAD OF THE INFLUENTIAL AAP vaccine advisory committee. Convinced that he had to do something quickly, he sprang into action, calling policymakers from the AAP, the CDC, and the National Vaccine Program Office (which coordinates activities of the FDA, the CDC, and the NIH). In late June and early July 1999, Neal Halsey conducted a series of teleconferences that would lead to one of the most momentous vaccine policy decisions ever made.

Few vaccine experts are more respected, more knowledgeable, or more dedicated than Neal Halsey. If he was concerned about something, people listened. "There is no question that it was Neal's concerns that drove a lot of this early on," recalled John Modlin, head of the vaccine advisory committee to the CDC. "He expressed a higher level of anxiety than the rest of us did. He wasn't convinced that there wasn't harm [caused by thimerosal]. And he was the driving force behind the AAP's decision." Halsey's initial proposal surprised his colleagues: he wanted to stop giving any vaccine that contained thimerosal. "We were vehemently opposed to that," recalled Jon Abramson, chairman of pediatrics at Wake Forest University School of Medicine and soon to be head of the AAP advisory committee. "Neal got into several heated discussions with someone at the CDC. I finally at one point said that this has got to stop. This screaming is getting us nowhere." "Neal really dominated the discussion," said Meg Fisher, another member of the AAP advisory committee. "He was adamant about what had to be done. And I think that most of the rest of the people on the call didn't feel that they knew the science well enough to have such a strong opinion and were uncomfortable with his opinion. But we didn't feel we had the ability to stop him. The first call set the scene. By the second call you had had time to read some of the science. And the more we read, the less excited we were to do anything. The more we felt that it was fine to say, 'OK, let's get thimerosal out of vaccines [eventually].' But Neal was just so adamant about it."

Although Halsey didn't represent the majority at the time, a confluence of events allowed him to prevail. First, Halsey pushed for a quick decision. "Most people don't realize the urgency with which Neal pushed this issue," recalled Georges Peter, former head of the National Vaccine Advisory Committee. "What started as a phone call to me on a Sunday night to the drafting of a statement was only about three weeks; most of that was within a ten-day period. And each call gave you the sense of a small snowball running down a mountain gaining in size and momentum to the point where it couldn't be stopped."

Also, three key decision makers weren't readily available: Walter Orenstein, head of the CDC's National Immunization Program; Jon Abramson, soon-to-be head of the AAP advisory committee; and John Modlin, head of the CDC advisory committee. Orenstein, attending his son's bar mitzvah in Israel, had to join calls at odd hours. Abramson, who was vacationing in the Canadian Rockies, joined when he could from telephone booths in remote areas. And in mid-May, just before the teleconferences started, John Modlin's boss—John Brooks, chairman of the Department of Pediatrics at Dartmouth Medical School— was almost killed in a car accident. "It was a tough period," recalled Modlin.

Typically, vaccine policies are made jointly by the CDC and AAP, occasionally with input from the National Vaccine Advisory Committee. But in this case, the perception that something needed to be done quickly, before these committees were scheduled to meet again, circumvented the process. Walter Orenstein later regretted the lack of vetting through these advisory committees. "To me one of the lessons is to put this through the usual decision-making process," he said. "What we did at the time was emergency decision making, and we had all sorts of groups involved that normally wouldn't have been." Georges Peter agreed. "The usual process by which an [AAP] statement is developed is that every member had to sign off on it before it went to the executive board," he said. "That process was bypassed. Neal was communicating directly with the executive board of the [AAP], and the statement was drafted. It might have been different if the [CDC advisory committee] had been able to weigh in."

The choice of teleconferences also helped to determine the outcome. "Had it not been conference phone calls, I don't think that the decision [to remove thimerosal quickly] would have been made," said Meg Fisher. "I think if we were all sitting in a room, someone would have gotten up and said, 'Neal, you're really off base here.' When you're face-to-face you can get the nuances." "The problem with teleconferences is that there's no chance to look around the room," said Georges Peter. "You could be talking to someone at a coffee break, and all of a sudden you realize, my gosh, there are six of us who all feel the same way." Larry Pickering, a member of the AAP advisory committee, also lamented the use of teleconferences. "Conference calls are fine for certain things," he said, "but not for major decisions. I personally feel that sitting in a room and looking someone in the eye and really being able to discuss things freely and openly are far superior and should be the way that major recommendations are made. Conference calls are not the way that this should have been done."

Despite the cauldron in which they found themselves, most people on the teleconferences gradually became convinced that nothing had to be done immediately. "A lot of the [AAP] committee didn't feel that this was a public health emergency," recalled Pickering, "and that we should move forward in a logical, systematic, well-thought-out way." "This was not a bomb that was immediately going to go off," recalled Peter. But Halsey didn't want to wait. On June 30, 1999, he called a meeting at the AAP executive offices in Washington, D.C. Before the meeting Halsey made it clear that he was going to ask doctors to use thimerosal-free vaccines for children whenever possible, to eliminate the birth dose of hepatitis B vaccine until thimerosal-free vaccines were available, and to stop immunizing children with influenza vaccines that contained thimerosal. Seeing the risk of potentially fatal infectious diseases as greater than the theoretical risk of thimerosal, most participants were surprised and angered by Halsey's recommendation. Martin Myers, deputy director of the National Vaccine Program Office, wrote a summary of the meeting. "Although the AAP is a major public health partner, there is great concern that the data do not support moving so precipitously," wrote Myers. "In addition, the manner by which the public is informed about the potential risk could greatly influence trust in our immunization programs."

Facing resistance from his colleagues, Halsey played his trump card. "If we weren't going to do what [Neal] wanted [to stop giving thimerosal-containing vaccines], we were told that he was going to do his own press release," recalled Myers. "It was very confrontational. We felt there was no alternative." Myers often thinks back to the events of June 1999. "I think it was necessary to tell the public that we had identified something that needed to be pursued further," he said. "[But] I think the policy decision we made was wrong. We scared a lot of parents. But I just don't know how we would have done it differently because Neal kept saying, 'Well, if you're not going to do it with me, I'm going to do it.' I don't know that we had a lot of options." Myers doesn't fault Halsey, whom he always considered to be acting in the best interests of children. "I've never questioned his motivations," said Myers. "He's a man who really cares about children. I just think he made a terrible mistake. He was convinced that he needed to do this and that he had to convince everyone else to do it."

Eventually, with the help of Surgeon General David Satcher, the group members reached a compromise. They wouldn't change their recommendations regarding thimerosal-containing DPT or Hib vaccines, but they would change them for thimerosal-containing hepatitis B vaccine. They reasoned that children born to mothers who weren't infected with hepatitis B virus were at a very low risk of getting infected. The solution was simple: delay the birth dose of hepatitis B virus in children whose mothers weren't infected with the virus. This would bring the level of mercury to which infants were exposed in the first six months of life below that deemed to be unsafe (for environmental mercury) by the EPA. It would, they reasoned, spare a public relations fiasco.

Both sides agreed to wait until after the July Fourth weekend to draft a consensus statement. On July 9, 1999, only three weeks after the FDA had held a meeting describing how much mercury was contained in vaccines, the AAP and the Public Health Service (PHS) issued a joint statement. The authors of the statement tried to be reassuring: "There are no data or evidence of any harm caused by the level of exposure [to mercury] that some children may have encountered in following the existing immunization schedule." Then the statement got to the heart of the matter, weighing relative risks. "The recognition that some children could be exposed to a cumulative level of [ethyl] mercury over the first six months of life that exceeds one of the federal safety guidelines on methylmercury now requires a weighing of two different risks. On the one hand, there is the known serious risk of disease and death caused by failure to immunize our infants. On the other hand, there is the unknown and probably much smaller risk, if any, of neurodevelopmental effects posed by exposure to thimerosal." The contradictory and confusing juxtaposition of phrases in this statement exposed the tensions among those who had written it. The risk to an infant from thimerosal was described as "probably much smaller" than the risk of disease from failing to vaccinate—the word *probably* implied that the risk from thimerosal might actually have been greater. But with the phrase *if any*, the risk from thimerosal was also described as potentially nonexistent. A press release issued by the AAP was even more confusing. "Parents should not worry about the safety of vaccines," it read. "The current levels of thimerosal will not hurt children, but reducing those levels will make safe vaccines even safer." Critics wondered how removing something that hadn't been shown to be unsafe could make vaccines safer. Doctors were also confused by the recommendation. But many parents reasoned that the only possible explanation for thimerosal's precipitous removal was that policymakers thought it was harmful.

One month after the joint statement, the National Vaccine Program Office (NVPO) held a meeting at NIH to talk about what had happened. Typically, the NVPO would have actively participated in setting vaccine policy. Now, they could only stand back and watch. Stan Music, an expert in toxicology and vaccine safety, recalled the meeting. "This was a very frustrating, even disappointing experience," wrote Music. "I guess I am not used to seeing people who are held up as the experts so thoroughly squander their potential for wise stewardship and lead the world of vaccines down an unnecessarily dangerous path. The rush for judgment and then for action was never addressed. Thimerosal has been maligned irreparably, even though there are no cases of thimerosal toxicity and all the risks are theoretical. Another blow—this time self-inflicted—has been delivered to vaccines." Vaccine researcher Stanley Plotkin also attended the NIH meeting. He compared the decision to delay the hepatitis B vaccine to the queen's court in *Alice in Wonderland*: "First the sentence, then the trial!" he said. Almost everyone agreed that the thimerosal decision had been the wrong one, a fact recorded in the meeting's minutes. But a transcript of the meeting was never released. Georges Peter later wondered why the outcome of this meeting was never published. "It became pretty clear [during the meeting] that the action that had been taken was precipitous and that you needed far more data before taking this position," he said. "I asked Marty [Myers, head of the NVPO] if we were going to publish the proceedings and he said, 'Yes.' But it never got off somebody's desk." As a consequence, few in the press or the public knew that after a review of the data under calmer conditions, most experts didn't agree with the plan to immediately remove thimerosal from vaccines.

In the wake of the thimerosal decision, CDC officials warned of a growing tide against vaccines. "In spite of the best effort in communication," they said, "the concept of good and bad vaccines may emerge." Acutely aware of the litigious climate in the United States, CDC officials further warned: "It is possible that many children born in the 1990s who have serious disorders of unknown etiology will now blame mercury in vaccines for their illnesses. This is particularly likely for illnesses that appear to be increasing, such as autism."

The CDC statement would be an ominous predictor of future events.

· · · ·

In October 1999, Neal Halsey appeared before members of the CDC's Advisory Committee on Immunization Practices, asking them to express a preference for thimerosal-free vaccines—in other words, to make it clear to the press and the public that the CDC felt that thimerosal might be unsafe. But the committee rejected Halsey's request, reasoning that all of the evidence supported the stance that thimerosal at the level contained in vaccines was safe. The committee members' vote in October 1999 showed that had they been involved from the beginning, a joint statement might never have been written, the hepatitis B birth dose might never have been delayed, and the public might never have been frightened by the theoretical risk of thimerosal. But by the time Neal Halsey appeared before the committee it was too late; events that would soon lead to distrust, litigation, and a handful of new and potentially dangerous therapies for autistic children had already been set in motion. "If I had known what would happen next," recalled Walter Orenstein, "I would never have gone along with Neal's plan."

Halsey could have learned from an event that had occurred several years earlier, one that had shown why it is virtually impossible to exercise the precautionary principle in the United States without significant collateral damage.

· · · ·

In 1961, a surgeon named Frank Gerow noticed that a plastic transfusion bag filled with blood felt like a human breast. Gerow reasoned that another substance, a gel called silicone, could be used to fill the plastic bag. So he visited representatives of Dow Corning, the leading manufacturer of silicone, and asked if they would be interested in designing a silicone-filled breast implant. (Silicone is made by stringing together silicon and oxygen, two of nature's most abundant elements. First synthesized before World War II, silicone has been used in artificial joints, heart valves, shunts, disposable needles, and syringes. As a consequence, most Americans have been exposed to silicone.) Gerow felt that the market for breast implants was limitless. Dow Corning agreed. One year later, in 1962, Frank Gerow performed the first silicone breast implantation.

Silicone breast implants became enormously popular, especially among Hollywood actresses and celebrities. Between 1979 and 1992, plastic surgeons put silicone breast implants into more than 100,000 women a year and breast implantation became the third most common cosmetic procedure in the United States, trailing only liposuction and eyelid surgery. A survey conducted by the American Society of Plastic and Reconstructive Surgeons found that 90 percent of women were satisfied with their implants and would choose to have them again.

But the tide was about to turn for breast implants.

In 1982, three women developed muscle aches, joint pain, fatigue, and weakness after receiving breast implants. All three women believed that breast implants had caused their symptoms. (Diseases such as rheumatoid arthritis, fibromyalgia, scleroderma, and lupus—all of which include symptoms related to joints and muscles—are collectively referred to as connective tissue diseases.) Later, an Australian physician reported their stories in the medical journal *Arthritis and Rheumatism*. Two years later, the lawsuits started. In 1984, a jury awarded $2 million to Maria Stern. Several years later, another jury awarded more than $7 million to Mariann Hopkins; it was one of the largest medical product awards in history.

In November 1990, breast implant litigants got their greatest gift: President George H. W. Bush appointed David Kessler to run the FDA. Kessler was a remarkable and brilliant man. Commuting between Boston and Chicago, he simultaneously received a medical degree from Harvard and a law degree from the University of Chicago, and his work schedule became one of legend. As FDA director, he labored tirelessly on behalf of the American public. "Caveat emptor [buyer beware] has never been—and never will be—the philosophy of the FDA," he said.

On April 16, 1992, amid growing concerns that silicone breast implants might be harmful, David Kessler banned them from the American market. At the time of the ban, no studies had examined the risk of connective tissue diseases in women who had received breast implants. Because studies hadn't been performed, Kessler had exercised the precautionary principle, removing breast implants from the market pending further study. He tried to reassure the public that breast implants hadn't been shown to be unsafe; he just wanted to study their safety before letting any more women use them. But his weak reassurances scared the public, causing many women who had implants to rush to have them removed. One woman, unable to afford the cost of another surgery, tried to remove her breast implants with a razor blade.

Kessler's removal of silicone breast implants from the market precipitated a flood of litigation. In one year, the number of lawsuits against Dow Corning increased from 200 to 30,000. Many of the lawsuits came from patient advocacy groups set up by lawyers to recruit clients. By 1994, breast implant manufacturers had been brought to their knees, forced to settle a class-action lawsuit for more than $4 billion, at the time the largest medical product settlement in American history. One billion dollars went to the lawyers. In May 1995, Dow Corning filed for bankruptcy.

On June 16, 1994, two months after breast implant manufacturers had agreed to settle the class-action lawsuit, Sherine Gabriel, a professor of medicine and epidemiology at the Mayo Clinic in Rochester, Minnesota, published a paper in the *New England Journal of Medicine*. Gabriel had evaluated 750 women who had received breast implants and compared them with 1,500 women who hadn't received them. The rates of connective tissue diseases were the same in both groups. Gabriel's conclusion was simple: "We found no association between breast implants and the connective tissue diseases and other disorders that were studied."

Gabriel's study was the first of many. In 1995, Jorge Sánchez-Guerrero, a researcher in the department of rheumatology and immunology at Harvard Medical School, examined the records of 90,000 women and published their findings, also in the *New England Journal of Medicine*. Again, women with breast implants were not more likely to have connective tissue diseases. Six more studies followed. Researchers at Wayne State University, the University of Calgary, the University of Kansas School of Medicine, the Johns Hopkins School of Medicine, the University of Pennsylvania School of Medicine, and the Harvard School of Public Health all agreed with Gabriel and Sánchez-Guerrero: breast implants didn't cause connective tissue diseases.

• • • •

THE MASSIVE SETTLEMENT OF THE CLASS-ACTION LAWSUIT DIDN'T put an end to breast implant litigation. Fifteen thousand women opted out of the settlement, choosing to seek their own awards in hopes of even greater gain. But there was one problem. When lawyers had settled the class-action suit, no studies had been performed to refute their claims. Now several researchers had published studies exonerating breast implants. When Neal Halsey evoked the precautionary principle in asking for the immediate removal of thimerosal from vaccines, he should have paid much closer attention to what followed in the breast implant story. Because the alliance among fringe scientists, personal-injury lawyers, and advocacy groups; the promotion of bad science; the cottage industry of unnecessary tests and lucrative consulting fees; and the accusation that the scientists who publicly had exonerated breast implants were part of a massive conspiracy funded by industry would all be repeated. The game plan had been set.

Two years after research failed to show a connection between breast implants and connective tissue diseases, a consortium of lawyers paid more than $1 million to a group called Fenton Communications to help them out. With the help of Fenton, a high-powered, well-connected public relations firm based in Washington, D.C., personal-injury lawyers pursued a multipronged strategy to counter the epidemiological studies that had absolved breast implants. First, Fenton promoted the work of an assistant professor of pathology at the UCLA Medical Center, Nir Kossovsky. Kossovsky had injected guinea pigs with silicone and found that it induced an immune response; he believed this explained why women with breast implants got connective tissue diseases. Kossovsky's argument contained a few holes: guinea pigs didn't develop connective tissue diseases; women with breast implants weren't at greater risk of connective tissue diseases; and no one had ever consistently found silicone-binding antibodies in people. Despite these differences, Kossovsky patented his observation, named his test Detectsil (a contraction of "detect silicone"), charged $350 for it, advertised it in magazines, founded a company called SBI Laboratories in Pittsburgh, and spent much of his time testifying on behalf of breast implant litigants.

Kossovsky wasn't the only one to set up a laboratory to perform such tests. Rahim Karjoo founded American Medical Diagnostics in West Covina, California, and Georgette Vojdani set up Immunosciences Lab in Beverly Hills. These two companies—devoted exclusively to detecting abnormal antibodies in women with breast implants—made millionaires of their founders. Douglas Shanklin and Douglas Smalley developed the "Silicon Sensitivity Test" or SILS, the flagship product of Memphis Pathology Laboratory in Memphis, Tennessee. And Scott Tenenbaum at Tulane University licensed his test, which sold for $200, to Autoimmune Technologies. Tenenbaum's diagnostic test—like all of the others—didn't detect silicone-binding antibodies because women with silicone breast implants didn't produce them. Still, Fenton promoted these scientists as having made important contributions to understanding how breast implants caused connective tissue diseases.

Later, the Institute of Medicine (IOM) reviewed hundreds of epidemiological and biological studies and concluded that breast implants didn't cause connective tissue diseases. But plaintiffs' lawyers and breast implant recipients claimed the IOM was part of a conspiracy to mislead the American public—a conspiracy financed by breast implant manufacturers. Plaintiffs' lawyers harassed Sherine Gabriel, the lead author on the first study exonerating breast implants. They wrote letters to the *New England Journal of Medicine* claiming she was biased and deeply embedded in the pocket of the manufacturing industry. They issued subpoenas, forcing Gabriel to comply with their wide-ranging and work-intensive demands. Gabriel later recalled the pressure of being the first to publish a paper that countered well-funded, highly motivated advocacy groups backed by personal-injury lawyers. "The magnitude of the demands [from personal-injury lawyers] was staggering," she said. "They wanted over eight hundred manuscripts from researchers that were here; they wanted hundreds of databases, dozens of file cabinets, and the entire medical records of all Olmstead County [Minnesota] women whether or not they were in the study. It has taken a huge amount of time and it has been extremely stressful. It has severely compromised my ability to do research." Which was, she suspected, the point. If plaintiffs' lawyers could tie up Sherine Gabriel, she would be less able to perform follow-up studies that further undermined the credibility of their position.

Gabriel wasn't the only investigator targeted by personal-injury lawyers. Marcia Angell, executive editor of the *New England Journal of Medicine* and author of *Science on Trial: The Clash of Medical Evidence and the Law in the Breast Implant Case*, also came under attack. "I was subpoenaed twice," said Angell. "They wanted a large number of documents that don't even exist, alleging contact between me and the manufacturers. They thought the manufacturers paid me."

Fenton's efforts to counter good science failed, and plaintiffs who had rejected the settlement watched their cases fall apart. On November 17, 2006, the FDA lifted its ban on silicone breast implants. An editorial in the *Wall Street Journal* added a post-script to the controversy: "While we're glad the FDA has over-turned fourteen years of politicized medicine by approving silicone breast implants, it's worth remembering the enormous price that has been paid: to the credibility of the legal system, in jobs lost, and in public health. And it's worth asking what is more toxic: the silicone implants preferred by thousands of women, or the trial bar that purports to 'protect' them." Despite the *Journal*'s protests, the temporary loss of silicone breast implants in the United States didn't exact much of a toll on human health.

For vaccines, however, the price would be much greater.

When public health agencies delayed the birth dose of hepatitis B vaccine, children suffered. Ten percent of hospitals, frightened by the notion of giving a thimerosal-containing vaccine to newborns, ignored recommendations and simply suspended the hepatitis B vaccine for *all* newborns. One three-month-old child born to a mother infected with hepatitis B virus in Michigan died of overwhelming infection, having failed to receive the vaccine in the nursery. Six babies in Philadelphia born to mothers infected with the virus also never got vaccinated, putting them at high risk of developing severe liver disease or liver cancer later in life. "I think stopping the birth dose of hepatitis B vaccine was an absolutely horrendous mistake," said Martin Myers. "It was one of the worst public policy decisions that we've ever made."

In retrospect, Neal Halsey and public health agencies hadn't exercised the precautionary principle—a principle that assumes that there is no harm in exercising caution. Rather, they had set sail on a grand experiment, believing that the harm caused by thimerosal was greater than the harm caused by disrupting the hepatitis B immunization program. One journalist, in an article for the *New York Times Magazine*, wrote, "If the autism trend begins to recede now that thimerosal has been removed, it could certainly suggest a cause. If it does decline, we might have Neal Halsey to thank."

The next few years would determine whether Neal Halsey's caution had paid off.

CHAPTER 5

Mercury Rising

Is the nation's spiraling rate of autism caused by mercury in vaccines? With over four thousand cases pending, a trillion dollars at stake, and public trust on the line, a firestorm is sweeping from the halls of science to the boardrooms of Big Pharma to the steps of the Capitol.

—SARAH BRIDGES

Lyn Redwood is a nurse practitioner who lives in Atlanta, Georgia. By the time her third child, Will, was born, she had been a medical professional for twenty years. "My son Will weighed in at close to nine pounds at birth," she said. "He was a happy baby who ate and slept well, smiled, cooed, walked, and talked all by one year of age." But after his first birthday, Will began to change. "He lost speech, eye contact, and suffered intermittent bouts of diarrhea, [then he was] diagnosed with pervasive developmental delay, a form of autism." When the AAP issued its press release in July 1999 urging the immediate removal of thimerosal from vaccines, Redwood called her doctor's office. "I reviewed [Will's] vaccine record and my worst fears were confirmed," she said. "All of his early vaccines that could have possibly contained thimerosal, had." Redwood believed she had found the cause of her son's autism.

Unlike Lyn Redwood, Sallie Bernard didn't have a background in health care. She's a businesswoman, owning and operating a market research company. Based in Cranford, New Jersey, the company manages focus groups and evaluates questionnaires. In September 1987, Bernard prematurely gave birth to triplets. One, Bill, weighed only three pounds and remained in the hospital for weeks, his development lagging far behind that of his brothers. It was Sallie Bernard's father-in-law who first noticed a problem. "I must tell you," he said, "I think something might be wrong with Bill. He's a little different than the other boys. He doesn't play with the same intensity they do." Bernard took her son to the doctor and was told he had dysphasia. That night, she looked the word up in the dictionary: *Dysphasia*: "an impairment or loss of speech or ability to understand language, caused by brain disease or injury." When Bernard saw the word *injury*, she wondered what could have injured her son. Later, she reached the same conclusion as Redwood. "Anyone familiar with the signs of mercury toxicity in children will recognize language difficulties [as a] common feature," she said.

Redwood and Bernard soon found each other. On July 3, 2000—almost one year to the day after public health officials had issued their warning about thimerosal—they submitted a paper to the journal *Medical Hypotheses* linking autism to mercury poisoning. Redwood and Bernard had scoured the medical literature looking for descriptions of children with autism or mercury poisoning; they wanted to see whether the two disorders were similar. Evaluating symptoms such as movement, speech, language, thinking, unusual behaviors, and psychiatric disturbances, Redwood and Bernard couldn't find anything that distinguished autism from mercury poisoning—the symptoms were exactly the same. They concluded that thimerosal "should be considered a probable source" of autism. Then they offered hope for a cure. If mercury caused autism, they reasoned, perhaps removing it from autistic children could help. "With one in one hundred and fifty children now diagnosed with autistic spectrum disorder," they wrote, "development of mercury-related treatments, such as chelation, could prove beneficial for this large and seemingly growing population." Chelation therapy for autistic children—the administration of chemicals designed to bind to mercury and to eliminate it from the body—was born. (The word *chelation* is derived from the Latin *chelos*, claw.)

The same year that Redwood and Bernard submitted their paper claiming mercury caused autism, they founded a parent advocacy group called Safe Minds, an acronym for Sensible Action for Ending Mercury-Induced Neurological Disorders. Their mission was to "end the health and personal devastations caused by the needless use of mercury in medicines." Redwood was angry that children had been and were continuing to be exposed to mercury, and she was angrier that the government hadn't seen her *Medical Hypotheses* paper as proof that mercury-containing vaccines were the problem. "We are in the midst of an autism epidemic and children diagnosed with learning disabilities continue to increase daily," she said. "The statement [by public health officials] that there is no evidence of harm does not equate with no harm having occurred." The phrase *no evidence of harm* would become an ironic manifesto for her cause. "The truth is that we have not adequately looked or we just refuse to see," challenged Redwood. "A recent national news article reported that some say we don't have a smoking gun. But the truth is the bullets are all over the floor."

• • • •

THE *NEW ENGLAND JOURNAL OF MEDICINE*, ONE OF THE MOST influential medical journals in the world, is read by more than 200,000 doctors and health professionals. On the other hand, *Medical Hypotheses*, where Redwood and Bernard had published their paper, has a circulation of about 200. No one in the medical community or the press pays much attention to publications in *Medical Hypotheses*. And although the media had followed Congressman Dan Burton's hearings on vaccines and autism, they didn't value his expertise, casting him as a lightweight with a personal agenda. If the thimerosal-causes-autism hypothesis was going to capture the attention of the American public, it needed a boost. On November 10, 2002, a journalist named Arthur Allen provided it. Allen wrote an article for the *New York Times Magazine* titled "The Not-So-Crackpot-Autism Theory." Allen had worked for the Associated Press as a foreign correspondent in El Salvador, Mexico, France, and Germany, tackling controversial issues unflinchingly. Later, he wrote a highly acclaimed book titled *Vaccine: The Controversial Story of Medicine's Greatest Lifesaver.* Although initially skeptical, Allen had been swayed by his interview with Neal Halsey. "Neal was a respected figure, and he knew what he was talking about," recalled Allen. "He said to me, 'You know, I've looked at a lot of things, and this is different.' That impressed me." The day after Allen's article appeared, National Public Radio (NPR) picked up the story.

Then the floodgates opened.

· · · ·

THE FIRST TO OFFER DATA TO SUPPORT REDWOOD'S AND BERnard's theories was a father-and-son team from suburban Washington, D.C., Mark and David Geier. Mark Geier had received his medical degree, and later a PhD in genetics, from George Washington University. For the next ten years he trained at NIH in Bethesda, Maryland, before becoming an assistant professor of obstetrics and gynecology at Johns Hopkins School of Medicine. His son, David, graduated from the University of Maryland, majoring in biology. In August 2002, two years after Redwood and Bernard submitted their paper to *Medical Hypotheses*, the Geiers submitted one to *Experimental Medicine and Biology.* In it, the Geiers wrote they "were initially highly skeptical that differences in the concentration of thimerosal in vaccines would have any effect on the incidence of neurodevelopmental disorders." But their skepticism turned to certainty. The Geiers examined data collected by the Vaccine Adverse Events Reporting System (VAERS), a federal program codirected by the FDA and CDC. Patients, doctors, nurses, or parents who feared a vaccine might have caused a dangerous side effect would fill out a one-page form and send it to VAERS; forms were then collected and analyzed. The Geiers studied reports of children after they had received DPT vaccines that did or didn't contain thimerosal. They reasoned that if the VAERS system received more reports of problems following thimerosal-containing DPT, then thimerosal must be the cause of the problems. The Geiers wrote, "An analysis of the VAERS database showed statistical increases in the incidence of autism, mental retardation, and speech disorders after thimerosal-containing [vaccines] in comparison with thimerosal-free vaccines." Where Lyn Redwood and Sallie Bernard had raised the hypothesis that thimerosal caused autism, Mark and David Geier had performed the first study that appeared to prove it.

In the spring of 2003, less than one year after they had published their VAERS study, the Geiers published another in the *Journal of American Physicians and Surgeons*. Again, using the VAERS database, the Geiers found the more mercury children received in vaccines, the more likely they were to develop autism and speech disorders; worse, they were also more likely to have heart attacks and epilepsy.

Evidence against thimerosal continued to mount.

In the summer of 2003, only a few months after they had found that thimerosal in vaccines caused heart attacks, the Geiers published another study in the *Journal of American Physicians and Surgeons*. To see whether mercury had caused autism, the Geiers fed dimercaptosuccinic acid (DMSA)—a chelating agent that binds to mercury—to 200 autistic children. The study also included eighteen children who didn't have autism. After administering the ninth dose, the Geiers collected urine and analyzed it. If mercury caused autism, the Geiers reasoned, autistic children should excrete more mercury in their urine after DMSA than nonautistic children. Their fears confirmed, the Geiers found "a strong, statistically significant association between greatly increased urinary mercury concentrations and the presence of autistic spectrum disorders in vaccinated children." (By "statistically significant," the Geiers meant that children weren't benefiting from their treatments randomly. Rather, the odds favored the Geiers' contention that improvement in symptoms was caused by chelation.) They concluded, "Data from this study increases the likelihood that mercury is one of the main factors leading to the large increase in the rate of autism."

• • • •

ALTHOUGH THE GEIERS HAD BEEN AMONG THE FIRST TO SUGGEST chelation therapy as a treatment for autism, it was a wealthy financier from northern California named J. B. Handley who made it a national movement.

In 2002, Handley's wife, Lisa, gave birth to a son, Jamie. For his first eighteen months Jamie was happy, playful, and engaging. Then he began a frightening descent into autism. He stopped making eye contact, stopped responding to his name, and spent days spinning around in circles. "We could have left the house and gone on a vacation to Hawaii and he would not have noticed," said Handley. "I cried six hours a day for a month. The reason we were devastated is because we believed what we were told about the prospects." But the Handleys refused to give up. Every day—at a cost of $5,000 a year—they rubbed a chelation chemical on Jamie's legs and forearms. The results were dramatic. "There's a light in his eyes now," said Lisa. "He laughs. It had been five months since we had seen this smile. He's a lot more cuddly and affectionate. He's seeking us out now. His emotions are back. He's happy." "Getting the mercury out is giving us our son back," said Handley.

By their estimation, the Handleys had witnessed a miracle. On May 24, 2005, J. B. Handley launched an organization called Generation Rescue. Its mission was simple: "Generation Rescue is a non-profit organization founded by parents of mercury-poisoned children dedicated to providing other parents with the truth about the cause of their children's neurological condition. We have united out of the shared bond, anguish, and outrage at discovering that our children have been mercury poisoned. Right now, thousands of parents armed with the truth are successfully healing their children." Handley reserved his highest praise for Mark and David Geier. "Motivated by truth," said Handley, "the Geiers have [published] numerous works proving the strong correlation between thimerosal and autism." Generation Rescue recruited more than 100 Rescue Angels—parents of autistic children who spread the word about the apparent miracle of chelation.

By advocating the use of mercury chelation for autistic children, Mark and David Geier were among the first to put Lyn Redwood's and Sallie Bernard's theory into practice. But they didn't stop with chelation. Based on the work of a well-known British researcher, the Geiers became the first to advocate a different, far more controversial therapy.

• • • •

SIMON BARON-COHEN IS A PROFESSOR OF DEVELOPMENTAL PSY-chology at Trinity College and the director of the Autism Research Center at the University of Cambridge. (He is also the first cousin of Sacha Baron Cohen, the English comedian and actor best known for his roles in *Borat* and *Da Ali G Show*.) In the late 1990s, Baron-Cohen published "Sex Differences in the Brain: Implications for Explaining Autism." Baron-Cohen knew that boys were far more likely to develop autism than girls. In his paper, he offered a possible explanation. He noted that boys and girls differed in the way that they responded to their environment: boys were more likely to be "systemizers" and girls "empathizers." Systemizers see their environment as a set of rules; to control it, they have to master the rules. Empathizers, on the other hand, see their environment as governed by other people's mental states; to control it, they have to respond with appropriate emotions. Baron-Cohen reasoned that boys had poorer social skills than girls because "other people's emotional states and behavior cannot easily be predicted and responded to using systemizing strategies." In support of his theory, he offered many studies showing that boys were more likely to play with mechanical toys

as children and, as adults, scored higher on physics and engineering problems. In contrast, girls scored higher on tests of
emotional recognition, social sensitivity, and verbal fluency; girls
also began to talk earlier than boys and were more likely to play
with dolls. Such differences are present almost immediately after birth; when one-day-old babies are shown a live face or a
mechanical mobile, boys prefer the mechanical object whereas
girls prefer the face. Interestingly, Baron-Cohen noted these differences also extended to animals. Young male monkeys prefer
to play with trucks, whereas young female monkeys prefer to
play with dolls. Baron-Cohen reasoned autism represented an
"extreme male brain." Autism was, in essence, extreme boyish-
ness.

Baron-Cohen also showed *why* boys approached their sur-
roundings differently than girls, even at a very early age. During
development in the womb, boys produced more of the hormone
testosterone than girls. Testosterone leads to changes in the structure of the male and female brain and ultimately to differences
in behavior. Mark and David Geier seized on this finding to
proffer their next treatment. If boys were more likely to become
autistic than girls, then perhaps autism could be treated by making them less boy-like. The Geiers proposed Lupron, a medicine
that shuts down testosterone synthesis.

In 2006, the Geiers wrote a paper that married the hypothesis of Lyn Redwood and Sallie Bernard with that of Simon
Baron-Cohen. They argued that because testosterone bound to
mercury, if children could be rid of testosterone, they could also
be rid of mercury. After injecting Lupron into more than 100
autistic children twice a day for weeks, the Geiers concluded
that it caused "a significant clinical amelioration in hyperactiv-
ity/impulsivity, aggression, self-injury, severe sexual behaviors,
and irritability behaviors that frequently accompany autistic
spectrum disorders." Mark and David Geier believed they had
found yet another cure for autism.

Excited about his new therapy, Mark Geier took to the air-
waves. On June 23, 2006, he appeared on Radio Liberty. "If you

look at [autistic] children," he said, "they have high testosterone, they masturbate at age six, they have mustaches, they're aggressive, and you can treat them by lowering their testosterone and removing the mercury." Geier believed he had discovered something about children with autism that had never been appreciated before: they were developing too quickly. He believed if he could slow sexual development, he could treat autism. Geier spoke rapidly and excitedly about his new therapy. "We've had unbelievable success," he said. "Virtually every one of the sixty or more children we've treated has improved in a very, very big way. And when you hear that kind of story you say, 'Well, there must be something wrong; it's too good to be true.'"

. . . .

MARK GEIER WASN'T THE ONLY SCIENTIST TO SUPPORT THE NOtion that mercury caused autism. Boyd Haley, a professor and chairman of the department of chemistry at the University of Kentucky, was a superb researcher who had published several papers in the *Proceedings of the National Academy of Sciences*—a publication of the National Academy of Sciences and a highly respected scientific journal internationally. Haley proposed that tubulin, a protein in cells necessary for their movement, was damaged by thimerosal. "Inhibit tubulin function with thimerosal injections," he said, "and you inhibit the immune response [causing autism]." Lyn Redwood was happy to have Boyd Haley on her side. "He understands the levels of exposure that our infants received," she said. "That's why Dr. Haley is such a wonderful advocate for us. He reads the science and understands it."

Soon Boyd Haley was joined by another biochemist: Richard Deth, a professor of biochemistry at Northeastern University in Boston. In 2004, Deth had written a paper published in *Molecular Psychiatry* that weighed in on the thimerosal debate. Using nerve cells grown in laboratory flasks, Deth had found that thimerosal inhibited an important metabolic pathway. He reasoned that some children were probably better able to use this pathway to excrete mercury, and to avoid its toxic effects, than others. And, like the Geiers, Deth believed he had found a cure. Testifying before Dan Burton's Committee on Government Reform, Deth said, "The good news that goes along with this [knowledge] is that metabolic interventions are proving to be effective for autism. These treatments include [vitamin] B_{12} itself, which can produce dramatic improvements in some kids. Giving B_{12} turns out to be an antidote for [mercury]." Richard Deth believed he had found a cause (alteration in cellular metabolism) and a treatment (vitamin B_{12}) for autism.

Later, Haley and Deth were joined by Mady Hornig, a researcher at the Mailman School of Public Health at Columbia University. Hornig had been influenced by the studies of Vijendra Singh, the Utah State University scientist who had proposed that autism was the result of autoimmunity. (Although Singh had testified before Dan Burton's committee that autoimmunity was caused by the MMR vaccine, not thimerosal.) Hornig chose to study inbred mice that had a very high rate of autoimmune diseases. She found that thimerosal stunted their growth, decreased their movement, and lessened their interest in exploring their environment. Using thimerosal, Mady Hornig believed she had made mice autistic.

Hornig's study was hailed by the press. On June 18, 2004, Sharyl Attkisson interviewed her on the *CBS Evening News with Dan Rather.* For several minutes Hornig explained how mice damaged by thimerosal provided a biological basis for understanding autism. The Internet site WebMD published a lengthy article titled "Mercury Linked to Autism Damage in Mice." And WNYW-TV in New York also featured the story of autistic mice. "[Mice] had changes in their brains that were reminiscent of some of the features that have been described as autism," said Hornig.

The case against thimerosal continued to build.

．．．．

WHEN NEAL HALSEY HELD THE SERIES OF TELECONFERENCES ON thimerosal in June 1999, Walter Orenstein (head of the CDC's

National Immunization Program) was away. When he returned and saw the uncertainty and confusion that had been created by changing immunization policy, Orenstein asked for a scientific study to examine the effects of thimerosal on children. He thought the study should be done using data from a surveillance program called the Vaccine Safety DataLink (VSD). Like Halsey, Orenstein didn't think that mercury in vaccines caused autism, but he wanted to make sure that it didn't cause other, more subtle neurological problems like speech or language delay, hyperactivity, attention deficit disorder, or tics.

Orenstein had chosen the VSD because of its unique properties. The VSD, which tracks patients belonging to several health maintenance organizations primarily in the west and southwest, contains a vast network of computerized patient information. Orenstein reasoned that CDC researchers could systematically check the VSD database and compare the amount of mercury children received in vaccines with their risk of various neurological problems. The idea made a lot of sense.

The researcher assigned the task of examining the records from the VSD was Thomas Verstraeten, an Epidemic Intelligence Service officer at the CDC. Using medical information from more than 100,000 children, Verstraeten mined the VSD database in an effort to determine whether mercury in vaccines had caused harm. After his first pass through the data, he concluded that it had. With the exception of autism, children who had received mercury in vaccines were more likely to have a variety of neurological problems. Verstraeten later presented his findings to Orenstein and others at the CDC. Concerned, Orenstein gathered fifty experts in the fields of autism, pediatrics, toxicology, neonatology, epidemiology, psychology, infectious diseases, and vaccines for a two-day meeting at Simpsonwood, a Methodist retreat in Norcross, Georgia. "Simpsonwood" would soon become a rallying cry for those who felt the government knew mercury had harmed children and done everything in its power to cover it up.

On June 7, 2000, Walter Orenstein called the meeting to order. "I want to thank all of you for coming here and taking time out of your busy schedules to spend the next day and a half with us," he said. Then he charged the group with its mission. "[For] those who don't know, initial concerns were raised last summer that mercury in vaccines might exceed safe levels. Analyses to date raise some concerns of a possible [relationship between] increasing levels of [mercury] in vaccines and certain neurological diseases."

Tom Verstraeten presented his data. He started with autism, concluding that the relationship between the amount of mercury in vaccines and the risk of developing the disorder was "not statistically significant." Then he talked about other neurological problems. He showed that children who had received mercury in vaccines were more likely to have tics, attention deficit disorder, and speech and language delays. The group knew what was at stake. If Verstraeten's preliminary data were right, vaccine makers, public health officials, and doctors had inadvertently poisoned a generation of children.

Paul Stehr-Green, an associate professor of epidemiology at the University of Washington School of Medicine and Public Health, was the first to see a flaw in Verstraeten's study. Stehr-Green reasoned that children who weren't getting vaccinated (and were therefore exposed to less mercury) might also be less likely to visit their doctor. "I think [this] impacts on [Verstraeten's] conclusions tremendously," he said. Stehr-Green was concerned that children who got more vaccines were more likely to be diagnosed with neurological problems not because they were actually at greater risk, but because they were more likely to come to the doctor. If this were true, then vaccinated children would also be more likely to be diagnosed with problems unrelated to mercury poisoning, like club feet. Stehr-Green postulated that if vaccinated and unvaccinated children visited their doctors with the same frequency, they might have the same risk of neurological problems. He asked Verstraeten to go back to the VSD database to determine if his hypothesis was correct.

Robert Davis, an associate professor of pediatrics and epidemiology at the University of Washington, was the next to show weaknesses in Verstraeten's study. Davis had been one of Verstraeten's coinvestigators. He knew the study was valid only if the computer records matched the medical records. (The VSD contains diagnoses listed by codes, not direct information from medical charts.) So he carefully compared the medical charts and computer records of 1,000 children. Davis was disappointed by what he found. "When we look at speech delay in particular," he said, "we find that, believe it or not, sometimes it is not even mentioned in the [medical] chart." Davis meant that administrators occasionally entered speech delay into the computer record when a child didn't have speech delay. Children with attention deficit disorder were also often coded incorrectly. "If attention deficit disorder is coded," said Davis, "you only have a 31 percent chance of finding a confirmed diagnosis of [it] in the medical record." Davis showed that Verstraeten had based his findings on inaccurate data; he realized investigators would now have to go back to every single medical record and make sure that it had been accurately recorded in the computer. This meant a lot more work and a lot more time to complete the study.

Tom Verstraeten and his coworkers took three years to address the problems that had been raised by the conferees at Simpsonwood. In November 2003, they published their findings in *Pediatrics*. Verstraeten had gone back to the medical records to verify the computer diagnoses; he had also addressed Paul Stehr-Green's question about possible differences between vaccinated and unvaccinated children in visits to the doctor. When he removed these confounding problems, Verstraeten found that his preliminary data had been misleading: mercury in vaccines did not cause harm. He concluded, "No consistent significant associations were found between thimerosal-containing vaccines and neurodevelopmental outcomes."

Orenstein had hoped that the VSD study would calm the fears of concerned parents. But some parents of autistic children were quick to see Simpsonwood as a cover-up: a secret meeting held out of sight of the press and the public so that the government could get its story straight. These parents had been convinced by Lyn Redwood and Sallie Bernard's mercury-causes-autism hypothesis, as well as by the studies of Mark and David Geier showing that vaccinated children were more likely to be autistic. They needed someone to counter Tom Verstraeten's new study, someone powerful enough to oppose the government's contention that vaccines didn't cause autism. They found him in one of America's most celebrated families.

• • • •

ROBERT F. KENNEDY JR., THE THIRD SON OF FORMER ATTORNEY general Robert F. Kennedy, worked for the Hudson RiverKeepers, a group dedicated to keeping the river free of dangerous pollutants. There, he developed an interest in mercury poisoning. In June 2005, Kennedy published an article in *Rolling Stone* magazine titled "Deadly Immunity." It began ominously. "In June 2000, a group of top government scientists and health officials gathered at a meeting at the isolated Simpsonwood conference center in Norcross, Georgia. Convened by the Centers for Disease Control and Prevention, the meeting was held to ensure complete secrecy. The agency had issued no public announcement of the session—only private invitations to fifty-two attendees. All of the scientific data under discussion, CDC officials repeatedly reminded the participants, was strictly embargoed." Unfortunately, Kennedy's article contained many errors. This was one of them. Verstraeten presented his findings at a public meeting at the CDC a few weeks after Simpsonwood.

Kennedy's article contained other inaccuracies. Kennedy wrote: (1) "[The CDC] withheld Verstraeten's findings, even though they had been slated for immediate publication, and told other scientists that his original data had been 'lost' and could not be replicated." Verstraeten published his study only after the problems with the preliminary data had been addressed. (2) "To thwart the Freedom of Information Act, [the CDC] handed over its giant database of vaccine records to a private company, declaring it off limits to researchers." Parents whose children's records were in the VSD had participated with the clear understanding that their medical information would remain confidential. (3) "Thimerosal appeared to be responsible for the dramatic increase in autism." During his preliminary investigation, Verstraeten had found that thimerosal might be responsible for a variety of neurological problems. Autism wasn't among them. (4) "By the time Verstraeten published his study in 2003, he had gone to work for GlaxoSmithKline, and reworked his data to bury the link between thimerosal and autism." Representatives from GlaxoSmithKline didn't influence Verstraeten's reanalysis; the fifty researchers at Simpsonwood did. (5) "In 1930, [Eli Lilly] tested thimerosal by administering it to twenty-two patients with terminal meningitis, all of whom died within weeks of being injected—a fact Lilly didn't bother to report in its study declaring thimerosal safe." Patients died from meningitis, not from mercury poisoning. (6) "By the age of six months [children] were being injected with levels of mercury one hundred eighty-seven times greater than the EPA's limit for daily exposure." Levels actually exceeded the EPA limit twofold, and that was for methylmercury, not thimerosal. (7) "Many of those on the CDC advisory committee who backed the additional vaccines [containing thimerosal] had close ties to the industry. Dr. Sam Katz, a paid consultant for most of the vaccine makers, shares a patent on a measles vaccine with Merck." Although Sam Katz was part of the team at Harvard that developed the first measles vaccine, he never patented it. (8) "Four of the eight CDC advisors who approved guidelines for a rotavirus vaccine laced with thimerosal had financial ties to the pharmaceutical companies developing different versions of the vaccine." No rotavirus vaccine has ever contained thimerosal.

Kennedy concluded his article with incendiary quotes, hoping to energize parents who felt the government had, in concert with pharmaceutical-company-contaminated scientists, pushed vaccines on an unsuspecting and now horribly damaged public. He quoted Boyd Haley: "You couldn't even construct a study that shows thimerosal is safe," said Haley. "It's just too darn toxic. If you inject thimerosal into an animal, its brain will

*Robert F. Kennedy Jr. was a passionate supporter of the notion that
thimerosal, a mercury-containing preservative once used in several vaccines,
caused autism (courtesy of Getty Images).*

sicken and swell. If you apply it to living tissue, the cells die. If
you put it in a Petri dish, the culture dies. Knowing these things,
it would be shocking if one could inject it into an infant without
causing damage." At the end of Kennedy's article, Mark Blaxill,
vice-president of Safe Minds, opened the curtain on the class-
action lawsuits that lay ahead: "The damage caused by vaccine
exposure is massive," he said. "It's bigger than asbestos, bigger
than tobacco, bigger than anything you've ever seen."

Flooded with letters and e-mails correcting Kennedy's many
mistakes, *Rolling Stone* issued a series of retractions on June 17,
22, and 24; but, given the number of national interviews that fol-

lowed, few in the media took notice. On June 20, 2005, Don Imus interviewed Kennedy on *Imus in the Morning*, a radio show that reached into the homes of millions of Americans. Imus came to the mercury debate through his wife, Deirdre, who, like Kennedy, is an environmental activist. "You see what autism does to people's lives," Kennedy told Imus. "It shatters the entire family. It destroys them. These kids [are] perfectly healthy, wonderful children, and they're brought into these pediatricians' offices, the wonderful family doctor who they trust. And they're given the shot because they're told its going to save your life. And they come out of that pediatrician's office and by late afternoon they are having seizures. Their parents bring them to the hospital and then the kids just disappear. This child who was a complete human being just disappears. He stops talking. He starts banging his head against the wall. He starts biting himself. He stops interacting with his siblings. And these people know this and they are continuing to inject children with this horrible, horrible toxin." The next day, Joe Scarborough interviewed Kennedy on MSNBC. "As it turns out," said Kennedy, "we are injecting our children with four hundred times the amount of mercury that the FDA or EPA considers safe." In his *Rolling Stone* article, Kennedy had overestimated the amount of mercury in childhood vaccines by ninetyfold; now he was off by two hundredfold.

. . . .

KENNEDY'S *ROLLING STONE* ARTICLE AND CONSEQUENT MEDIA tour helped galvanize parents; no longer would they have to stand alone in their fight against public health agencies.

More help was on the way.

In 2005, the same year Kennedy published his article in *Rolling Stone*, a journalist named David Kirby wrote *Evidence of Harm: Mercury in Vaccines and the Autism Epidemic: A Medical Controversy*. Kirby's book sounded many of the themes that had been raised by Kennedy, touting the research of Boyd Haley, Richard Deth, Mady Hornig, and Mark and David Geier; trumpeting the apparent wonders of chelation; and disparaging the epidemiologic study done by Tom Verstraeten and the CDC. "The CDC has been unable to definitively prove or disprove the theory that thimerosal causes autism," wrote Kirby. Picking up on the phrase used by Lyn Redwood in an earlier statement to the press, Kirby wrote, "But 'no evidence of harm' is not the same as proof of safety. No evidence of harm is not a definitive answer. And this is a story that cries out for answers." Kirby, like Kennedy, believed public health officials had ignored the public's health. "A small group of parents, aided by a handful of scientists, physicians, politicians, and legal activists," wrote Kirby, "has spent the past five years searching for answers. Despite heavy resistance from the powerful health lobby, these parents never abandoned their ambition to prove that mercury in vaccines is what pushed their children, most of them boys, into the hellish, lost world of autism." Kirby, like Kennedy, also believed the CDC was involved in a conspiracy to hurt children, critically altering data presented by Verstraeten at Simpsonwood and refusing to talk to him about it. "Many of the public health officials who discount the thimerosal theory were unwilling to be interviewed for the book," he wrote.

Kirby's book was an instant success. Don Imus interviewed him several times on *Imus in the Morning*, calling *Evidence of Harm* a "wonderfully researched book that looks at both sides of the issue," and he bemoaned powerful lobbies working against the health of American children. "These big huge chemical companies," said Imus, "they pay too much money to all these politicians, and their huge lobbying groups, with the pharmaceutical and chemical companies, [and] they're not interested in knowing the root cause." Kirby also lamented the conspiracy: "The CDC wouldn't talk to me," he said to Imus. Joe Scarborough interviewed Kirby on MSNBC. "We're pumping babies full of mercury in these vaccines," said Scarborough. "We are," Kirby agreed. Tim Russert interviewed Kirby on *Meet the Press*, and Montel Williams interviewed him on *The Montel Williams Show*. Kirby was thrilled with the media attention. To readers on an Internet bulletin board, he wrote, "We are now getting about five media requests a day, mostly from people we never even pitched to. It's amazing. Other columnists are mentioning the book in their pieces. Today, the NPR show *To the Point* called and tentatively booked me for Friday. Local radio and smaller city papers are all over this, believe me. And from the media not ONE NEGATIVE WORD so far. Not one. Only from the Quackwatch bloggers, the pediatricians, the National Network for Immunization Information, the CDC, and for some reason, the State of Minnesota. AND, drum roll, please, *Evidence of Harm* will be mentioned in the *New York Times* three times before the month ends."

Robert F. Kennedy Jr. and David Kirby had taken the debate about whether mercury caused autism and made it personal. They likened the CDC to the fox guarding the henhouse, recommending vaccines while at the same time evaluating their safety. And when Tom Verstraeten left the CDC for GlaxoSmithKline, they accused him of having fronted for the pharmaceutical industry all along. Rather than guarding the public's health, the CDC was simply watching out for its own health. Many picked up on the conspiracy theme proposed by Kennedy and Kirby. Sallie Bernard said that scientists who claimed that vaccines were safe had "an interest in not finding a connection." Dan Burton said: "I'm so ticked off about my grandson. And to think that the public-health people have been circling the wagons to cover up the facts! Why, it just makes me want to vomit!" "This is the biggest cover-up in medical history," said Mark Geier. "It's bigger than 9/11 and AIDS and no one knows about it."

• • • •

POLITICIANS ALSO JOINED THE FRAY.

On May 31, 2005, Joe Lieberman, a senator from Connecticut, appeared on Imus's radio show. "If you look at the statistics about the incredible increase in autism during the '90s," said Lieberman, "it increased 4,000 percent. And at the same time we changed the requirements for the normal vaccine dosage, including a lot of the vaccines that had thimerosal in them. You've got to ask yourself, isn't there a connection between these two things?" Like Kennedy, Kirby, and Imus, Lieberman was unwilling to accept reassurances from the CDC. "I don't care how many respected institutions are on the other side," he said.

On June 22, 2005, Christopher Dodd, another senator from Connecticut and soon-to-be presidential candidate, also appeared on Imus's show. "I read the Bobby Kennedy piece [in *Rolling Stone*], which is pretty good," he said. Dodd was upset that U.S. manufacturers were shipping autism-causing vaccines to other countries. "We have stockpiles of this stuff," said Dodd, "[and] we just ship it to the rest of the world, have them use it. It's bad enough there. God forbid you're living in some third-world country and your child ends up with this problem."

On March 20, 2006, John Kerry, a senator from Massachusetts and the Democratic candidate for president of the United States in 2004, added his support to the growing list of politicians who believed vaccines caused autism. "Let me tell you an amazing story," he began when he too appeared on Imus's show. "During the early days of Katrina, UPS [and] FedEx put together a couple of planes to fly stuff down [to New Orleans]. And the UPS driver I rode with just started to talk. He tells me he's got twins and one of the twins got sick. The doctor says [to him], 'Well, we've got to give him a vaccination.' They give him the vaccination while the kid is still sick, and within days this kid starts changing and showing symptoms. And that child who was vaccinated has autism today. And this family is struggling with it. And they believe as deeply as they can, it's the thimerosal that caused this reaction. You and Deirdre have been terrific on this issue. And yet, we still have mercury in vaccinations around the country. It's absurd. I don't get it. We ought to stop."

Perhaps the most influential champion of parents who believed mercury had damaged their children was Dave Weldon, a physician and congressman from Florida. Weldon vigorously defended Mark and David Geier in their quest to take a closer look at the data generated by Tom Verstraeten, and he chastised

John Kerry (left), *like Robert F. Kennedy Jr., also used his public stature to trumpet the notion that vaccines caused autism (courtesy of Getty Images).*

the Institute of Medicine for its failure to support and encourage researchers like Richard Deth and Mady Hornig. "I am very disturbed," said Weldon, "by the continuing number of reports I receive from researchers regarding their experiences. It is past time that individuals are persecuted for asking questions about vaccine safety." Weldon claimed that scientists who had found vaccines were unsafe had lost grants and had had difficulty getting their findings published. "Others," said Weldon, "report overt discouragement, intimidation, and threats, and have abandoned the field of research. Some have had their clinical privileges revoked and others have been hounded out of their institutions." Weldon drew an analogy between mercury researchers and Andrew Wakefield: "When researchers find things that are unpopular, that goes against the party line, they are punished for it."

Weldon wanted autism research to be directed by Safe Minds, the advocacy group founded by Lyn Redwood and Sallie Bernard.

In May 2004, he told Julie Gerberding, the director of the CDC, to stop focusing on epidemiological studies like the one done by Tom Verstraeten (which showed that mercury in vaccines didn't cause autism) and to start focusing on laboratory studies, like the ones done by Richard Deth and Mady Hornig (which showed it did). "Fine," said Gerberding. "Can your office put together something? Can you get me a wish list of all the kinds of protocols that you would like to see funded?" Weldon's chief of staff, Stuart Burns, called Lyn Redwood. "Hey Lyn," he said, "do you think Safe Minds could put together a list of research projects on autism and vaccines you would like to see the federal government pursue?" No longer would research be determined solely by the NIH, the CDC, and physicians in academic institutions with an expertise in autism; rather, it would be directed by a parents' advocacy group.

On August 26, 2004, Arnold Schwarzenegger, governor of California, trumped his fellow politicians by banning mercury-containing influenza vaccines from his state. (In 2004, only multidose influenza vaccines contained thimerosal as a preservative.) By April 2006, six states had followed his lead; in 2007, another seventeen states were considering similar bans. Because only limited supplies of thimerosal-free influenza vaccines are available, public health officials worried that banning thimerosal was equivalent to banning influenza vaccines for some children, putting them at risk of severe and occasionally fatal disease.

• • • •

AFTER DAVID KIRBY wrote EVIDENCE OF HARM, HE MADE A PREdiction: "If the number of three- to five-year-olds [with autism] in the California [health care] system has not declined by 2007," he said, "that would deal a severe blow to the autism-thimerosal hypothesis." Kirby knew that thimerosal had been removed from vaccines routinely given to young infants in the spring of 2001. And he knew that six years would be plenty of time to figure out whether thimerosal had caused autism. Kirby wasn't the only person who recognized the importance of the natural experiment that was about to unfold. On July 8, 2000, Sharon Humiston, the mother of an autistic child and a physician in Rochester, New York, appeared before Dan Burton's committee. "Please don't miss the opportunity to study the results of removing thimerosal from vaccines," said Humiston. "As the manufacturers change to mercury-free formulations, I hope someone is doing a definitive study to see if autism rates plummet." "The final verdict," said Boyd Haley, "will come with observing the rate of autism now that thimerosal has been removed from the infant vaccine program. The truth will come out." Mark and David Geier were more certain of the outcome. "The first step in the process is the immediate removal of thimerosal from all vaccines," they said, "which we predict will result in the end of the autism epidemic."

In April 2004, only three years after thimerosal had been removed from vaccines routinely given to young infants, public health officials in California reported a small drop in the rates of autism. It was the first such drop in twenty years. Lyn Redwood was cautiously optimistic. "This information does give us hope," she said, "but we will monitor future data to be sure this is in fact a real trend downward versus a one-time anomaly. Since it would take five years [from the removal of thimerosal from vaccines] to see results, this new data is right on track with our expectations." David Kirby was also excited by the drop in rates: "Is it too early to tell if this is a permanent and meaningful trend? Of course. Could there be other explanations for the drop, such as a budget-crunching reduction in services? Perhaps. But this very decline, at this very moment, has long been predicted by supporters of the thimerosal-autism theory. At the very least, the quivers of their detractors have been emptied of one arrow."

In the spring of 2006, Mark and David Geier published the results of their own investigation of autism rates, now that thimerosal had been removed, in the *Journal of American Physicians and Surgeons*. "We predicted the rates of autism would begin to decrease," wrote the Geiers. "Thimerosal started to be removed in July 1999. What we are seeing is decreasing trends." Richard Deth, the biochemistry professor who had shown that thimerosal damaged laboratory cells, said, "This study is exactly the kind of thing people have been waiting for."

. . . .

THE THIMEROSAL-CAUSES-AUTISM STORY WAS SO COMPELLING that Hollywood decided to make a movie about it. In 2006, Participant Productions, having previously made *Syriana*, *North Country*, *An Inconvenient Truth*, *Fast Food Nation*, *Murderball*, and *Good Night and Good Luck*, optioned the rights to David Kirby's *Evidence of Harm*, immediately promoting the film on its Web site: "When their children descend into the frightening world of autism a group of parents discover a disturbing link between thimerosal, a mercury-based preservative found in vaccines, and the steady rise in autism. One tenacious mother, Lyn Redwood, risks her family to battle the FDA, CDC, and the American government, despite efforts from pharmaceutical companies and government officials to suppress evidence and prevent parents from gaining restitution for their children's conditions." Kirby was thrilled that his book was going to be made into a movie. "I thought we might want to put politics aside for a moment to say 'Hooray for Hollywood'," Kirby declared on his Web site. "I spoke last night with *Evidence of Harm* producer Ross Bell, who produced *Fight Club*, by the way. [Bell] told me that the first draft of the screenplay is beginning to take shape. Once that is finished, the script will go to Participant Productions for notes and review, and I will get to see a copy as well!! Once the director is hired, and I have a few personal favorites I am rooting for, then the whole project 'goes public.' Bottom line, look for a MAJOR announcement in the media sometime before the winter is over. Even just the ANNOUNCEMENT of this movie should make some waves. I am so grateful to Participant Productions. I know that they have been contacted by people who are 'extremely concerned' that this movie is being made. I bet they are!" Kirby had his choice for the part of Lyn Redwood. "How about Toni Collette for the female lead?" he wrote. "She does 'frantic mom' pretty damn well."

• • • •

LYN REDWOOD AND SALLIE BERNARD HAD PROPOSED THAT MER-cury in vaccines caused autism. Mark and David Geier had per-formed studies apparently showing that autistic children could be treated successfully with mercury-binding therapies. Richard Deth and Mady Hornig, using laboratory cells and experimen-tal mice, believed they had shown why mercury caused autism. And now, with mercury out of vaccines, the rates of autism ap-peared to be declining. So dramatic was the evidence against vaccines that a major, well-respected production company was going to make a movie about it. Everything was coming to-gether. Everything made sense.

But the next few years would reveal that it was all a mirage.

CHAPTER 6

Mercury Falling

Falsehood flies, and the truth comes limping after; so that when men come to be undeceived, it is too late; the jest is over, and the tale has had its effect.

—JONATHAN SWIFT

Walter Orenstein had been the first to propose a study evaluating the risk of thimerosal. But because the preliminary results of the study were horribly flawed—as discussed by the conferees at Simpsonwood—it hung under a cloud. Lyn Redwood, Sallie Bernard, Mark and David Geier, Boyd Haley, Robert F. Kennedy Jr., David Kirby, and everyone else wedded to the thimerosal-causes-autism hypothesis roundly dismissed the study's final conclusions. But Orenstein's suggested study wasn't the only one to examine whether thimerosal caused harm. Eight more followed.

In August 2003, Paul Stehr-Green published a paper in the *American Journal of Preventive Medicine*. Stehr-Green studied children with autism in Sweden and Denmark from the mid-1980s through the late 1990s. He found the risk of autism increased *after* thimerosal had been removed from vaccines. By the late 1990s, when health officials had completely eliminated thimerosal, the number of children with autism was higher than it had ever been, exactly the opposite of what would have been expected if thimerosal caused autism. Stehr-Green concluded, "The body of existing data is not consistent with the hypothesis that increased exposure to thimerosal-containing vaccines is responsible for the apparent increase in the rates of autism."

In September 2003, Kreesten Madsen, an epidemiologist from the University of Aarhus in Denmark, published a paper in *Pediatrics*. Madsen examined the medical records of 1,000 children diagnosed with autism between 1971 and 2000. Like Stehr-Green, he found that between 1992 and 2000, after thimerosal had been removed from vaccines in Denmark, the incidence of autism skyrocketed. Madsen concluded the data "did not support a correlation between thimerosal-containing vaccines and the incidence of autism."

In October 2003, Anders Hviid, an investigator from the Danish Epidemiology Science Center in Copenhagen, published a paper in the *Journal of the American Medical Association*. Hviid studied the records of Danish children between 1990 and 1996, during which time thimerosal had been removed from vaccines. Like Madsen and Stehr-Green, Hviid found the number of children with autism increased after thimerosal had been eliminated. He concluded, "The results do not support a causal relationship between childhood vaccination with thimerosal-containing vaccines and development of autistic spectrum disorder."

One year later, in September 2004, Jon Heron, an epidemiologist from the University of Bristol in the United Kingdom, published a study in *Pediatrics*. Heron examined the records of 14,000 children who had received different amounts of thimerosal in vaccines between 1991 and 1992. He wanted to see if he could find a relationship between the amount of thimerosal babies had received and the risk of neurological problems. He did. The more thimerosal children received, the *less* likely they were to be hyperactive or to have difficulties with hearing, movement, or speech. Heron, like Stehr-Green, Madsen, and Hviid before him, had found exactly the opposite of what parents concerned about thimerosal would have expected. He concluded, "We could find no convincing evidence that early exposure to thimerosal had any deleterious effect on neurological or psychological outcome."

The same month Jon Heron published his study, Nick Andrews, an epidemiologist from the Communicable Disease Surveillance Center in London, also published a study in *Pediatrics*. Andrews examined the records of 100,000 children who had received different amounts of thimerosal. Like Heron, Nick Andrews found the more thimerosal children received, the less likely they were to develop neurological problems like attention deficit disorder. Again, the amount of mercury in vaccines didn't presage the development of autism. Andrews concluded, "There was no evidence that thimerosal exposure via vaccines caused neurodevelopmental disorders."

In 2004, after reviewing more than 200 epidemiological and biological studies of the relationship between thimerosal and autism, a committee of the Institute of Medicine (IOM) issued a statement to the press and the public: "The committee concludes that the body of epidemiological evidence favors rejection of a causal relationship between thimerosal-containing vaccines and autism. The committee further finds that potential biological mechanisms for vaccine-induced autism that have been generated to date are theoretical only." The institute recommended that autism research dollars should be spent on more fruitful leads.

Two years later, Eric Fombonne, an epidemiologist at McGill University in Montreal, put another nail in the thimerosal-causes-autism coffin. Fombonne surveyed 28,000 children born between 1987 and 1998. During that time, babies could have received vaccines that contained anywhere from zero to more than 200 micrograms of thimerosal. Fombonne found that the group with the highest risk of autism had received no mercury. He concluded, "The findings ruled out an association between [autism] and high levels of thimerosal exposure comparable with those experienced in the United States." Like researchers in Denmark, Canada, the United Kingdom, and the United States,

Fombonne had found that the number of children diagnosed with autism in Canada had increased throughout the mid- to late 1990s. This increase occurred at the same time that thimerosal had been removed from vaccines. Obviously, removing thimerosal hadn't caused the increase. But what had? Fombonne had an explanation: "Factors accounting for the increase include a broadening of diagnostic concepts and criteria, increased awareness and, therefore, better identification of children with [autism] and improved access to services." In other words, Fombonne reasoned that there wasn't an epidemic of autism; rather, broadening the definition of the disability to include mildly affected children, as well as heightened awareness among parents and doctors, had accounted for the increase.

In September 2007, Bill Thompson at the CDC published the most comprehensive and definitive study to date. Thompson carefully determined the exact amount of mercury that 1,000 children had received in vaccines and performed more than forty separate neurological and psychological tests. The study, which was published in the *New England Journal of Medicine*, took four years to complete. Its results were consistent with those of the other six: vaccines containing mercury hadn't caused harm.

Finally, in January 2008, Robert Schechter and Judy Grether from California's Department of Public Health took a closer look at the rates of autism from 1995—six years before thimerosal had been removed from vaccines—to 2007, six years after it had been removed. They found what everybody else had found: the rates of autism continued to increase. In an accompanying editorial titled "Thimerosal Disappears but Autism Remains," Eric Fombonne wrote, "Parents of autistic children should be reassured that autism in their child did not occur through immunizations."

• • • •

THOSE WEDDED TO THE NOTION THAT THIMEROSAL CAUSED AUtism were furious. They declared the epidemiological studies

meaningless and were sickened that the IOM had been persuaded by them. What really mattered, they said, were the biological studies, like those of Richard Deth and his laboratory
cells and Mady Hornig and her mice. On NBC's *Meet the Press,*
David Kirby, now a spokesman for the cause, said: "You need to
look at the biology, the toxicology; you need to look at the cellular level. Virtually half of the evidence that was presented [to
the IOM] against the theory [that thimerosal caused autism]
was epidemiological. The other half supporting the theory was
largely biological. And yet the committee gave a preponderance
of emphasis to the epidemiological evidence and rather, I would
say, gave short shrift to the biological evidence."

Kirby had underestimated the power of epidemiological
studies.

• • • •

ALTHOUGH VACCINES HAVE PROBABLY SAVED MORE LIVES THAN
any other medical intervention, they have come with a price—
occasionally causing severe, even fatal, side effects. Epidemiological studies have been the single most powerful tool to show
that vaccines, like all medicines, are imperfect.

In 1998, the FDA licensed a rotavirus vaccine, and the CDC
recommended it for all infants. The vaccine, designed to prevent
a common cause of fever, vomiting, and diarrhea, had been tested
in 10,000 babies before licensure. But after it had been given to 1
million babies, the vaccine was found to be a rare cause of intestinal blockage called intussusception. The problem wasn't trivial. Babies with intussusception can die when bacteria from the
intestine enter their bloodstream, or they can die when blood
vessels in the intestine are damaged, causing massive bleeding.
Epidemiological studies showed rotavirus vaccine caused intussusception in about one of every 10,000 babies who got it. Of
the 1 million children who had received this vaccine, 100 suffered intussusception, and one died. Within months of the vaccine's release, the CDC had discovered the problem and withdrew
its recommendation—a testament to CDC diligence and post-

licensure surveillance. The CDC's quick and decisive response to the problem with the first rotavirus vaccine belied accusations that it couldn't be trusted to determine whether vaccines were safe.

The rotavirus vaccine experience was just one example of the power of epidemiological studies. Natural measles infection occasionally causes the number of platelets (cells in the bloodstream necessary for clotting) to decrease. The disorder, called thrombocytopenia, can be quite serious, occasionally causing life-threatening bleeding. Measles vaccine also can cause temporary thrombocytopenia, albeit rarely, in about one of every 25,000 vaccinated children. Investigators unearthed this rare problem with measles vaccine using a series of powerful epidemiological studies. In 1976, public health officials in the United States, fearing that an unusual outbreak of influenza among soldiers at Fort Dix (New Jersey) signaled the start of the next influenza pandemic, immunized millions of Americans with what was called the swine flu vaccine. Unfortunately, some people immunized with the vaccine developed a rare form of paralysis called Guillain-Barré Syndrome. Epidemiological studies showed that the vaccine was the cause. One of every 100,000 people who got swine flu vaccine—400 people among 40 million—had been afflicted.

Problems caused by vaccines as rare as one in 10,000, one in 25,000, or one in 100,000 have been readily detected by epidemiological studies. If autism, a disease that affects one of every 150 American children, was caused by thimerosal, epidemiological studies would have detected it. Indeed, even if thimerosal in vaccines accounted for only 1 percent of autism—one in 15,000 children—epidemiological studies would have found it. Instead, after examining the records of hundreds of thousands of children, investigators working in both North America and Europe couldn't find any evidence of a relationship between thimerosal and autism. It wasn't that their studies were poorly designed or that they had been part of a vast international conspiracy to hide the truth. They couldn't find a relationship because it wasn't there to be found.

• • • •

When David Kirby had pleaded with Tim Russert on *Meet the Press* to dismiss epidemiological studies in favor of laboratory studies, he was asking for science-in-reverse. Typically, scientists determine that something is a problem (by doing epidemiological studies) *before* determining why it is a problem (by studying its effects in the laboratory). That's because studies in animals and cells can be quite misleading. For example, in the 1950s, researchers found that hamsters injected with SV40—a monkey virus that had inadvertently contaminated early lots of polio vaccine—developed large tumors under their skin. Some of these tumors were so big they weighed twice as much as the hamsters. But many epidemiological studies performed during the past fifty years have clearly shown that SV40 virus doesn't cause cancer in people. Studies in hamsters, although frightening, weren't predictive. In the 1960s, researchers found that hamsters injected with adenovirus—a virus that causes colds, pneumonia, and bronchitis in people—got cancer. If these hamster studies were revealing, then people infected with adenovirus should be at higher risk for cancer. But they're not. In the 1970s, researchers showed that laboratory cells exposed to large amounts of formaldehyde became cancerous. But people who work with formaldehyde, like morticians who use it to preserve dead bodies, aren't at greater risk of cancer. And everyone has small amounts of formaldehyde circulating in the bloodstream, the result of a natural process called single-carbon metabolism.

In each of these cases, biological studies of animals didn't predict what was happening in people. Laboratory studies can also work the other way; instead of sounding a false alarm, they have been falsely reassuring. For example, in the 1950s, researchers were interested in making a vaccine to prevent polio. The vaccine, pioneered by Jonas Salk, was made by inactivating polio virus with formaldehyde. After it was licensed, five companies stepped forward to make it. One company, Cutter Laboratories of Berkeley, California, made it badly. Because Cutter hadn't completely inactivated its vaccine, more than 100,000 children were inadvertently injected with live, dangerous polio virus. Seventy thousand got mild polio, 200 were permanently paralyzed, and ten were killed. It was one of the worst biological disasters in American history. Before releasing its vaccine, Cutter had tested it extensively in cells, mice, and monkeys to make sure it didn't contain live polio virus. But the laboratory studies had been falsely reassuring. Edwin Lennette, the director of the Virus Laboratory for the State of California and one of the country's leading virologists, stated the obvious: "The only way to determine whether something is a problem in people is to test it in people."

A more recent example can be found in Merck's AIDS vaccine trial, suspended in November 2007. The vaccine, which had been remarkably effective in mice and monkeys, failed miserably when tested in people. Echoing the words of Edwin Lennette, Peggy Johnston, director of the AIDS Vaccine Program at NIH, said, "Mice lie, monkeys sometimes lie, and humans never lie."

Cigarette smoking is another example of how David Kirby's plea for science-in-reverse can be misleading. In 1939, Alton Ochsner, a cancer surgeon in Louisiana, was the first to propose that cigarette smoking caused lung cancer. Researchers tried to prove Ochsner's theory in laboratory animals, but results were inconclusive. Studies in people told a different story. In the early 1950s, two epidemiological studies, one published in the *Journal of the American Medical Association* and the other in the *British Medical Journal*, clearly showed that people who smoked cigarettes were at greater risk of lung cancer—and the more they smoked, the greater the risk. Tobacco industry representatives refused to believe it, reasoning that because epidemiological studies were only "statistical," they didn't prove anything. The truth, they claimed, lay in laboratory studies (which had been inconclusive), not in epidemiological studies (which had been damning). But Bradford Hill, the lead investigator on the British study, disagreed: "In this particular problem, what experiment can one make? We may subject mice or other laboratory animals to such an atmosphere of tobacco smoke that they can—like the old man in the fairy story—neither sleep nor slumber. And lung cancer may or may not develop to a significant degree. What then? We may have strengthened the evidence, but we must, I believe, invariably return to man for the final proof." Further epidemiological studies consistently showed that although lung cancer caused by cigarette smoking was rare, affecting less than 1 percent of those who smoked, it was real. The results of these epidemiological studies no longer allowed an industry that wished to believe smoking didn't cause cancer to hide behind laboratory studies that had proved worthless.

. . . .

EPIDEMIOLOGICAL STUDIES AREN'T THE ONLY EVIDENCE THAT EX-onerates mercury as a cause of autism. Understanding how mercury moves through the environment also revealed why thimerosal is an unlikely culprit.

Mercury is part of the earth's surface, released into the environment by burning coal, rock erosion, and volcanoes. After it is released, it settles onto the surface of lakes, rivers, and oceans where it is converted by bacteria to methylmercury. Methylmercury is everywhere—in the fish we eat, the water we drink, and the infant formula and breast milk we feed our babies. There is no avoiding mercury.

Because everyone drinks water, everyone has small amounts of methylmercury in their blood, urine, and hair. A typical breast-fed child will ingest almost 400 micrograms of methylmercury during the first six months of life. That's more than twice the amount of mercury than was ever contained in all vaccines combined. And because the type of mercury in breast milk (methylmercury) is excreted from the body much more slowly than that contained in vaccines (ethylmercury), breast milk mercury is much more likely to accumulate. This doesn't mean that breast milk is dangerous, or that infant formula is dangerous, or that water is dangerous. Not at all. It means only that anyone who lives on the planet will consume small amounts of mercury all the time. During legislative hearings to ban mercury-containing vaccines, some politicians have stood up and said: "I have zero tolerance for mercury." This kind of statement makes for a great sound bite. But because mercury is an inescapable part of our environment, politicians with zero tolerance for it are going to have to move to a different planet.

Studies of mercury in vaccines showed why the amount of a substance is important when determining whether it is harmful. But mercury isn't the only example of this. The earth contains other heavy metals such as lead, cadmium, arsenic, beryllium, and chromium. Large amounts of any of these metals can be toxic, but very small amounts of them are found in most people. The quantity factor also works the other way: large amounts of nontoxic substances can be harmful. For example, sometimes people working in the sun drink gallons of water during the day to replace the fluids they've lost while sweating. But if the water doesn't also contain the minerals they've lost (like sodium), then minerals in the blood can be dangerously depleted, causing seizures. This doesn't mean water is toxic to the nervous system. It only means that if large amounts of water are ingested quickly, it can be dangerous. (Water intoxication drew national attention on January 12, 2007, when Jennifer Lea Strange, a mother of three, died after drinking two gallons of water as part of a radio promotion in Sacramento, California, called "Hold Your Wee for Wii." Two years earlier, a Chico State student died of water intoxication during a hazing.) The first person to recognize the relationship between dose and danger was the chemist Paracelsus, who early in the sixteenth century declared, "The dose makes the poison."

The thimerosal-causes-autism theory didn't make sense for other reasons. Lyn Redwood and Sallie Bernard had argued that symptoms of mercury poisoning were indistinguishable from symptoms of autism. But Karin Nelson, a neurologist from NIH, and Margaret Bauman, a neurologist from Harvard Medical School, showed these two disorders were in fact quite different. Children with mercury poisoning suffer narrowing of their field of vision, whereas children with autism don't have visual problems. Children with mercury poisoning can become severely psychotic, whereas children with autism, although socially aloof, aren't psychotic. Children with mercury poisoning have heads that are smaller than normal, whereas children with autism have heads that are larger than normal. Nelson and Bauman concluded, "The typical clinical signs of [mercury poisoning] are not similar to the typical signs of autism." Consistent with the findings of Nelson and Bauman, children in Minamata Bay who had suffered neurological damage from mercury poisoning didn't have a greater risk of autism.

．．．．

IN THE FACE OF OVERWHELMING EVIDENCE ABSOLVING MERCURY in vaccines, one parent advocacy group did what its predecessor had done during the breast implant litigation: it hired Fenton Communications to manage the media.

On June 8, 2005, after five studies exonerating thimerosal had been published, J. B. Handley's Generation Rescue hired Fenton to design a full-page advertisement to be published in the *New York Times*. The ad, which cost more than $150,000, bore a title in bold, one-inch-high letters: "MERCURY POISONING AND AUTISM: IT ISN'T A COINCIDENCE." The ad featured several quotes from prominent politicians. Senator John Kerry: "Mercury has been linked to autism." Congressman Dan Burton: "Numerous scientists have testified and presented credible, peer-reviewed research studies that indicated a direct link between exposure to mercury and autism." Congressman Dave Weldon: "Mercury is a neurotoxin. And, in the 1990s, children, infants, and unborn children were exposed to significant amounts of mercury at the most critical point of their development." And Senator Joe Lieberman: "I think parents have a just cause, and I don't care how many respected institutions are on the other side. This is a fight worth fighting." One year later, on April 6, 2006, Generation Rescue hired Fenton to design another full-page advertisement to be published in *USA Today*. The ad, which cost

$100,000, was boldly titled "IF YOU CAUSED A 6,000% IN-
CREASE IN AUTISM WOULDN'T YOU TRY TO COVER IT
UP, TOO? Under the headline was a quote from Robert F. Kennedy Jr.: "It's time for the CDC to come clean with the American public."

• • • •

WHILE SOME PARENTS WAGED WAR ON STUDIES EXONERATING
vaccines by placing advertisements in newspapers, others tried
to limit their impact by threatening the scientists who had performed them, the journalists who believed them, and the public
health officials who trumpeted them.

Sarah Parker is an assistant professor of pediatrics at the University of Colorado. After she wrote an article criticizing the
Geiers' epidemiological studies—and later appeared on a local
radio program to debate David Kirby—she received a series of
threatening e-mails and phone calls. "They would say that you
need to retract this paper immediately, or else," recalled Parker.
"They would say, 'Don't you have children of your own?' in a way
that was just very aggressive, very frightening." Fearing for the
safety of her eighteen-month-old daughter, Parker called the police
to obtain a restraining order on one particular caller. "My impres-
sion after talking to the police [was that the callers] seemed to
be very well trained in not using the words that would get them
in trouble with the law. So, they wouldn't make a direct threat,
saying, 'I'm going to hurt you' or 'I'm going to hurt your children.'
They wouldn't tell me what the 'or else' was. [But] I was worried
that they were going to do something to my family. I quit an-
swering my phone for about a year." The threats worked. Sarah
Parker never published another paper on vaccine safety. "I'm interested in vaccine safety," she says, "but I'm not sure how much
I would want to publish [and have to] live with those weekly
threats."

The IOM was also a target. Marie McCormick, a professor
of maternal and child health at Harvard's School of Public
Health, was the head of the committee that found no relationship

between vaccines and autism. "We were expecting trouble," said McCormick. "There had been e-mail threats concerning one of the speakers and one of the staff so that we actually changed the venue of the meeting [to] an auditorium where the committee could leave the meeting room without going through the audience. The committee was urged to stay in the same hotel. They all came on a single bus and were driven into the garage under the [National Academy of Sciences] building and went directly to the meeting room. Given the level of the threats, security was clearly beefed up. And one of the committee members resigned in what he described as a low probability–high [impact] threat."

On December 29, 2003, Kim Strassel, a member of the editorial board of the *Wall Street Journal*, wrote about the politics of autism. The editorial mentioned that science had begun to refute the notion that thimerosal caused autism. Strassel remembered what happened next. "A woman called our office," she said. "She was very pleasant. But her goal wasn't to talk to me about the editorial. It wasn't to try to convince me or argue with me. It was to confirm that I was the person who had written it. The next thing I knew my name and my contact details had been published on the Internet and circulated to these parents groups. They wanted to put my name and details out so that there could be a direct assault on a person. They want people. They want names and faces." Soon, Strassel was besieged with threatening e-mails and phone calls. "There were a couple [of e-mails] that suggested that if I had a child, I better hope that I took good care of him. I didn't have children then. [But] it's the most awful threat that someone can make to someone else. I figured I could look after myself. But you leave your kids. You drop them off places. They're not under your constant supervision. It's a very scary threat." Strassel is circumspect about the experience. "I'm in the business of editorial writing," she said. "Editorial writing inspires great passions among people, and if you don't have a thick skin, then you shouldn't be in this business. I'm very used to people telling me that I'm wrong or that I'm stupid or unethical. But I've never had a situation where people claimed that I was killing their children, where people were suggesting that I was part of some grand conspiracy between politicians and corporations to ruin their lives. And I never had anyone suggest to me that I should worry about my own family."

The CDC also received a series of threatening e-mails. One stated, "Forgiveness is between you and God. It is my job to arrange a meeting." Another lamented, "I'd like to know how you people sleep straight in bed at night knowing all the lies you tell and the lives you know full well you destroy with the poisons you push." The CDC contacted the FBI, instructed its staff on safety issues, hired more security guards, and showed employees how to respond if pies were thrown in their faces. "It's like nothing I've ever seen before," said Melinda Wharton of the CDC's National Immunization Program.

Threats from parents weren't limited to the United States. At the peak of the MMR-autism controversy in England, David Salisbury, the director of immunization for the Department of Health, suffered a similar incident. "I had a letter that came to work that said 'If I ever see you on television defending vaccines, I will come and find you.'"

• • • •

ALTHOUGH SOME PARENTS HAVE BEEN SKEPTICAL OF THE SCIEN-tists and public health officials who failed to find that vaccines caused autism, questioning their motives and occasionally threatening them, they haven't been similarly skeptical of the vast array of autism therapies, all of which are claimed to work and all of which are based on theories that are ill-founded, poorly conceived, contradictory, or disproved.

Dan Burton's Committee on Government Reform and conferences hosted by groups such as Autism One, Defeat Autism Now, and the MIND Institute have provided a venue for those who offer the latest and best cure for autism. For example, Vijendra Singh claimed vaccines caused an immune response against the sheath that lines nerve cells, arguing that drugs like steroids, which decrease inflammation, could treat autism. (Autistic children don't have inflammation of their nerve sheaths.) Mary Megson claimed cod liver oil, which contains vitamin A, extended the field of vision of autistic children. (Autistic children don't have a restriction of their visual fields.) John Upledger claimed spinal fluid was getting trapped in the brains of autistic children and causing increased pressure. He said he could relieve the pressure by cranial manipulation. (Autistic children don't have increased pressure on their brains.) Mark and David Geier claimed autistic children had more mercury in their bodies than nonautistic children and getting rid of it by chelation therapy could help. (Autistic children don't have excess mercury in their bodies; cells damaged by heavy metals aren't healed by chelation, and no well-performed study has ever shown chelation treats autism.) Stephen B. Edelson, director of the Edelson Center for Environmental and Preventive Medicine, claimed he could treat autistic children with high-intensity sound waves, calling his miraculous new therapy sonar depuration. (Sonar depuration therapy has never been formally tested.) Edelson said sonar depuration helped damaged brain cells to regenerate: "A classic example is you can take a six-year-old, remove half their brain, and within two years the child will be perfectly normal." (Children who lose half their brains don't grow them back.) Paul Harch, president of the International Hyperbaric Medical Association, said autistic children weren't getting enough oxygen to their brains and that hyperbaric oxygen yielded dramatic results. Now he was ready to test his revolutionary new therapy. "Before I get started, I wanted to make an announcement," said Harch. "The International Hyperbaric Medical Association and the American Board of Clinical Medical Toxicology as well as the Oklahoma University Health Science Center and School of Medicine are going to conduct the first evidence-based medicine study on the only two effective therapies that have been identified for autism: hyperbaric oxygen therapy and chelation therapy." (If Harch was about to perform the first study of these two therapies, how did he know they were effective?) "This is the only study that will address two of the major underlying problems with the majority of autism cases; and we're going to prove it in the next three years." (Three years have passed.)

It gets worse. Rashid Buttar, an osteopathic physician in North Carolina, prescribes a chelation medication called TD-DMPS, which is rubbed on the skin. TD-DMPS isn't licensed by the FDA for use in children, has never been formally tested in children, and has never been shown to cause the excretion of heavy metals. Buttar, who also injects children with filtered urine, is now the subject of a disciplinary action. He isn't alone in such quackery. Another doctor offers "RNA drops" at $100 a bottle. (RNA is completely degraded by enzymes the moment it touches the skin or tongue.) The "body ecology diet" proffers fermented foods, such as kimchee, sauerkraut, and kefir. Others recommend camel's milk; "foot-soaking machines"; laser therapy; bacteria-containing nasal sprays; pig whipworm eggs; and baths in magnetic clay, claiming the small black flecks left in the bathwater are tiny bits of metal pulled from the body. And for parents willing to travel to Mexico, Costa Rica, or China, the promise of stem-cell transplantation awaits.

But by far the most extensive network of physicians offering alternative therapies for autism belongs to Defeat Autism Now (DAN), a group based in San Diego and part of the Autism Research Institute. Like Andrew Wakefield, DAN practitioners believe autism is caused by toxic substances that enter the body through a leaky gut. (Studies have failed to prove that autistic children have leaky guts, and brain-damaging toxins have never been identified.) Their three-pronged approach to treating autism includes removing substances that could damage the gut, reinoculating the gut with healing bacteria, and repairing the gut with nutrients. At DAN conferences, held twice a year on the East and West Coasts, this therapeutic approach is referred to as "the three R's": remove, reinoculate, and repair.

Before DAN therapies can begin, children must undergo a daunting array of tests, which may include a complete blood count; hemoglobin, hematocrit, and liver function tests; urinary function tests; or blood tests of ammonia, organic acids, iron, amino acids, metals, and metallothioneins. Additional tests may include urinary analyses for specific peptides; stool analyses to detect yeast, harmful bacteria, parasites, fat, chymotrypsin, acid, and meat and vegetable fibers; and blood tests to detect food allergies and previous infection with viruses such as measles, mumps, rubella, Epstein-Barr virus (mononucleosis), and cytomegalovirus.

As in the case of laboratory tests, autism therapies are diverse, expensive, and unproven. To eliminate yeast overgrowth, autistic children may receive fluconazole, nystatin, and amphotericin B; to remove parasites, sporanox; to reduce harmful bacteria like clostridia, metronidazole. To reinoculate the gut, children may swallow microorganisms such as *Lactobacillus acidophilus*, *Lactobacillus rhamaonse*, *Sacchromyces boulardis*, *Bifidobacter bifidum*, and *Streptococcus thermophilus*. Finally, to restore the gut, children may be fed or injected with various combinations of vitamins (B_1, B_6, B_{12}, C, and E); minerals (magnesium, zinc, and calcium); folate; amino acids; fatty acids and ketoglutaric acid; di- and trimethylglycine; taurine; melatonin; immune globulins; creatine; digestive enzymes; glutathione; carnitine; carnosine; cod liver oil; activated charcoal; or colostrum (human milk).

But of all the alternative therapies proposed by DAN practitioners, none is more pervasive than diets free of casein (dairy products) and gluten (wheat, barley, and rye). These diets were first proposed in the early 1980s by a Norwegian biochemist, Kalle Reichelt. Reichelt believed that he had found abnormal protein fragments (called peptides) in the urine of patients with autism and schizophrenia—peptides that resulted from the incomplete breakdown of grains and dairy products. Reichelt's findings didn't stand the test of time. Several other groups of scientists, using sophisticated techniques like high-performance liquid chromatography, have consistently failed to find what Reichelt found. Unfortunately Reichelt's elimination diets haven't been eliminated. And they've come with a price. A recent study by NIH and Cincinnati Children's Hospital Medical Center found that autistic children deprived of calcium and vitamin D in dairy products developed osteoporosis, a dangerous thinning of the bones.

Tests and treatments recommended by DAN are often expensive, costing thousands and sometimes tens of thousands of dollars. And because the tests are seldom covered by medical insurance, parents pay for them out-of-pocket. Some tests, such as those for stool analysis, urinary peptides, organic acids, and amino acids, are often performed only by special laboratories working in concert with DAN. Because many physicians don't believe DAN tests and treatments are useful, the Autism Research Institute lists on its Web site the names and locations of clinicians who do, calling them DAN doctors. It's a cottage industry of false hope.

About 300 DAN doctors practice in the United States. Although they often tout their treatments as harmless, they may not be. Vitamin B_6 injections can damage nerves, and excess vitamin A can damage the liver and cause a buildup of pressure on the brain. Many DAN doctors have been disciplined by their state medical boards for practicing medicine unethically or illegally, and several have had their medical licenses suspended or revoked. Two DAN doctors have been censured for injecting hydrogen peroxide intravenously.

Despite the poor foundation on which alternative therapies are built, some parents claim dramatic results. How is this possible? How is it possible that so many different therapies based on so many different theories could work? The answer lies in the study performed by researchers in North Carolina who determined whether secretin—the intestinal hormone first touted by Victoria Beck—treated autism. Autistic children injected intravenously with secretin were judged by their parents to have improved, but so were children who had been injected with salt water. This didn't mean that both secretin and salt water treated autism; it meant that parents who had participated in the secretin study had a strong desire to see medicines make their children better. Parents not only had an emotional investment in the idea that secretin worked, they also had a financial investment, some having spent thousands of dollars to obtain the drug. Secretin has now fallen out of favor. But chelation, Lupron, sonar depuration, cranial manipulation, laser therapy, camel's milk, magnetic clay, and hyperbaric oxygen are no different. Parents want to believe these therapies work because they desperately want their children to get better; they don't want to watch them struggle anymore. But if all of these therapies are ever carefully tested, they will likely meet the same fate as secretin.

. . . .

WHILE SOME PARENTS HAVE CONTINUED TO BELIEVE IN ALTERNAtive therapies, others haven't. Sharon Humiston, a doctor from Rochester, New York, told Dan Burton's committee, "The worst day of my life was not the day the developmental pediatrician told me, almost apologetically, that my son [Quinn] was autistic. The worst day came later when the specialists told me that Quinn's progress was minimal after a year and a half of intense [behavioral] therapy, and that this made his prognosis grave. My family's response was very typical. We reached out and embraced a succession of therapies, each touted as a lifesaver. We heard that gluten allergy was the cause, and we changed Quinn's diet. Later, we tried a phenol-free diet, megavitamins, anti-yeast medications, and cranio-sacral massage. Each therapy was supposed to get at the cause of Quinn's autism. Each was expensive, and each was a failure."

Jim Laidler, a physician, wrote an article describing the seduction of alternative therapies: "I consider myself to be a very scientific person. While growing up, I was skeptical and inquiring and naturally gravitated to the sciences. My first brushes with pseudoscience and quackery in medical school left me convinced that it would never happen to me. I was wrong. A year or so after my [younger] son was diagnosed with autism—with no hope for a cure in sight—I was feeling desperate for anything that might help him. My wife attended a conference about biological treatments for autism. She came back extremely excited, having heard story after story about hopeless cases of autism cured by a variety of simple treatments. I was initially skeptical, but my desperation soon got the better of me. We started out with simple therapies—vitamins and minerals—but soon moved on to the 'hard stuff': the gluten- and casein-free diet, secretin, and chelation. Some of it seemed to work—for a while—and that just spurred us to try the next therapy on the horizon. I was hooked on hope, which is more addictive and dangerous than any street drug. [But] the beginning of the end was when my wife—suspecting that some of the supplements we were giving our older son [who also has autism] weren't having any effect—stopped them all without telling me. I saw no difference, even after two months, when she finally told me. We had been chasing our tails, increasing this and decreasing that in response to every change in his behavior—all the while his ups and downs had just been random fluctuations. My eyes began to open. The final step in my awakening came during a Disneyland vacation. My younger son was still on a gluten- and casein-free diet, which we swore had been a significant factor in his improvement. We had lugged at least forty pounds of special food on the plane with us. In an unwatched moment, he snatched a waffle and ate it. We watched with horror and awaited the dramatic deterioration of his condition that the 'experts' told us would inevitably occur. The results were astounding—absolutely nothing happened. I began to expect that I had been very foolish. In the following months, we stopped every treatment except speech and occupational therapy for both boys. They did not deteriorate and, in fact, continued to improve at the same rate as before—or faster. Our bank balance improved and the circles under our eyes started to fade. And, quite frankly, I began to get mad at myself for being so gullible and for misleading other parents of autistic children. Looking back on my experiences with alternative autism therapies, they seem almost unreal. Utter nonsense treated like scientific data, people nodding in sage agreement with blatant contradictions, and theories made out of thin air and unrelated facts—and all of it happening right here and now, not in some book. Real people are being deceived and hurt, and there won't be a happy ending unless enough of us get together and write one."

• • • •

IN JANUARY 2008—SIX YEARS AFTER THIMEROSAL HAD BEEN RE-moved from vaccines given to young infants in the United States—health officials from California announced the state's most recent rates of autism. Because children with autism were typically diagnosed between three and five years of age, six years of thimerosal-free vaccines should have been enough time to see whether the rates of autism had decreased. They hadn't. The short-lived trend downward in 2004 had been falsely inter-preted. In fact, the rates were increasing—dramatically. The grand experiment had been performed, and the results were clear. Rob-ert Davis, then head of the Immunization Safety Office at the CDC, stated the obvious: "If you remove cars from the high-ways, you'll see a marked decrease in auto-related deaths. If thimerosal was a strong driver of autism rates and you remove it from vaccines, [researchers] would have seen some sort of decline—and they didn't."

• • • •

WHEN DAVID KIRBY WROTE *EVIDENCE OF HARM*, HE CLAIMED TO be a journalist interested in getting at the truth. He said that he didn't have a dog in the hunt; as a journalist, he couldn't. Now Kirby faced eight epidemiological studies that disproved the no-tion that thimerosal caused harm and the clear and simple fact that removal of thimerosal from vaccines hadn't even slowed the number of children being diagnosed with autism. With all of this evidence in hand, many leading national and international organizations interested in the health and well-being of children declared vaccines hadn't caused autism. Reasonably, Kirby should have taken a step back, looked at the data, and declared his hypothesis disproved. Other journalists had done exactly that. In his *New York Times Magazine* article, Arthur Allen was one of the first journalists to alert the public about the possibility that vaccines caused autism. But as Allen read the epidemiological studies and followed the California autism rates, he gradually became convinced thimerosal didn't cause autism. "The data were just really clear," he said.

Kirby, on the other hand, refused to admit defeat. On November 11, 2006, he descended into parody. Before an audience of parents of children with autism in San Diego, Kirby, during a debate with Arthur Allen, offered another explanation. Obviously, vaccines weren't the only problem. Kirby claimed that children were getting autism because China was burning more coal, causing plumes of mercury-filled smoke to envelop the West Coast; to prove it he showed slides labeled "Shanghai Plume" and "Mongolian Plume." Kirby also noted an increase in California wildfires which, he reasoned, had caused more mercury to be released into the environment. But Chinese coal and forest fires weren't the only problems. Kirby also believed autism in California had been caused by an increase in cremations, resulting in the release of mercury from dental fillings. A few months later, the CDC announced that the rates of autism in the United States were the highest they had ever been. Kirby had tried to explain an increase in California autism rates using his Chinese smoke–California wildfires–cremation hypothesis. But he was hard-pressed to use this same hypothesis to explain why national rates of autism were increasing. Arthur Allen remembered the debate: "He just came up with this wild stuff. In this case a plume of mercury-containing vapor from China had descended on southern California. How can you debate that? It's just patently ridiculous."

In June 2007, Participant Productions, facing overwhelming evidence that vaccines didn't cause autism, stopped production on the movie version of *Evidence of Harm.*

. . . .

KIRBY'S PERSISTENT PORTRAYAL OF AUTISTIC CHILDREN AS VACCINE-damaged angered some parents. They didn't see autistic children as poisoned by anything and were disappointed by parents who had paraded their children at marches on the Capitol or at protests in front of the CDC with T-shirts reading "Damaged by Mercury."

Perhaps the most passionate and persuasive defender of autistic children from the damaged-by-mercury label is Kathleen Seidel from Peterborough, New Hampshire: a librarian, author, and creator of the Web site neurodiversity.com. "I get really angry at those who work parents up into a nasty emotional state," said Seidel. "I just think it's destructive to be led down this path of endless recriminations that goes with being encouraged to think that kids are autistic because they're damaged." In response to what Seidel described as the "rampant hostility and cynicism and suspicion and conspiracy theorizing that I saw happening on the *Evidence of Harm* listserv," on May 29, 2005, she sent a letter to David Kirby, urging him to see the effect he has had: "[I want] to encourage you to consider how it feels for an autistic person to hear incessant, gruesome, emotionally charged de-

Demonstrators protest the use of mercury in vaccines during a march on the U.S. Capitol, July 20, 2005 (courtesy of Getty Images).

scriptions of autism by non-autistic individuals who regard autism as an unmitigated tragedy, as completely unacceptable;
descriptions that insist that an autistic person's experience of the
world is a consequence of poisoning, and whose cognitive and
behavioral peculiarities are worthy of total eradication. I hope
that you will consider that when you make public statements
about autism you are presuming to speak about an entire class
of people who are autistic for life, not simply the subset of parents and minor children with whom you've become acquainted
over the past couple of years. For every parent eager to 'recover'
their child and 'lose the diagnosis,' there are autistic citizens
who will always have the diagnosis and will always wear the
label; and who are affected by the manner in which that label is
bandied about by those who hate what it represents to them.
People too often talk about 'autism' as if it is separate from autistic people. It is not. And those autistic people and their families—
autistic children, the autistic-spectrum adults whose very exis-
tence you have questioned, and families like mine—will still be
around, still dealing with the stigma of 'contamination' that you
have helped to promulgate, long after the royalties dry up and
you have sailed off to your next journalistic destination."

In the end, it was Kathleen Seidel who pulled back the curtain, revealing untold motives and unsavory relationships among
those who had trumpeted the notion that mercury caused autism—
and exposing a much darker place.

CHAPTER 7

Behind the Mercury Curtain

The greatest defeat of all would be to live without courage,
for that would hardly be living at all.

—GERALD R. FORD JR.

Kathleen Seidel lives on the edge of a cow pasture in rural New Hampshire, in a small, yellow raised ranch house stuffed with books and music. Her ready kindness, joyous laugh, and instinctive warmth belie a relentless, unsympathetic determination to expose the doctors, scientists, journalists, and politicians who have promoted the notion that vaccines cause autism. "I have a big fraud button," she says. "And it gets pushed every now and then."

Seidel was raised in Anaheim, California. Her father was a chemical engineer, her mother a music and special education teacher who worked with severely impaired children. In 1973, after graduating from high school, Seidel attended the University of California at Santa Cruz, majoring in English and Russian literature and Book Arts and later venturing to New York City, where she earned a master's degree in Library Science from Columbia University. After three years as a children's librarian in Asheville, North Carolina, Seidel returned to New York City where she met her future husband, a guitar player. She worked for

Project Orbis, a flying ophthalmologic surgical teaching hospital, and the Taconic Foundation, where she was the assistant to the director. "It was a great job," she recalled. "These people were wealthy and privileged, but they were doing philanthropic work in housing and youth employment in consonance with their conscience and principles. They were good people. I learned that nobody needs to apologize for living on Park Avenue."

In 1995, Seidel and her husband moved to New Hampshire to raise their family. Five years later, her life changed. "We got our diagnosis [of autism] in the family in 2000," she recalled. "The first thing that really woke us up was Oliver Sacks's article, "An Anthropologist on Mars." (Sacks had recounted the story of an autistic woman named Temple Grandin, who, as a professor at Colorado State University, had become a world-renowned designer of humane livestock facilities. The title of the essay describes how Grandin felt in social situations.) Initially, Seidel had trouble connecting with other parents of autistic children online, getting more out of talking to them in local support groups: "I joined one or two mailing lists, [but] I ended up not participating too much because it just wasn't the kind of support I needed. You say 'autism,' but there are many different levels of impairment. I just didn't connect with a lot of the conversations, especially with the preoccupation with vaccines. I didn't feel the need to find an external cause because by that time I'd realized the apple doesn't fall far from the tree."

Seidel decided to start a Web site and blog, calling it neurodiversity.com, the domain name coming from a family brainstorming session in 2001. "My first vision of it was as a library to organize some of the chaos of information that I had encountered out there," she recalled. "Now it's up to about three hundred different subject categories. Over time it's kind of bifurcated into half library and half soapbox." Seidel became increasingly concerned by some parents' obsession with vaccines. "In the spring of 2004, I was really blown away by how more and more people were talking about vaccines," she recalled. "In addition to the newfound ferocity of the rhetoric surrounding the whole vaccine thing, I was really turned off by some of the catastrophic language that people used to talk about autism because that just increases parents' stress when they're looking for information."

In late 2004, Seidel's Web site changed course. "I got an e-mail from a friend of mine saying, 'Get a load of this,'" she recalled. "[My friend] linked me to a report of a talk given by Boyd Haley at a meeting of a group called Doctors for Disaster Preparedness. He was talking about mercury. [Haley, a professor of chemistry at the University of Kentucky, had been the first to publicly propose a mechanism for how thimerosal might cause autism.] At one point in his talk he said, 'So one cow drops dead from mad cow disease and the government spends millions of dollars trying to figure out what's going on. Well, here we've got this epidemic of mad child disease and they're not doing a thing.' So I thought, you've got to be kidding me. How dare you make up a phrase like that to talk about my kid? Would you like your kid to be likened to an animal, to a mad animal? Within twenty-four hours I had a petition up on my Web site. And I got about three hundred signatures in the first few days. Then I thought that if [Haley] had any confidence in [his] science, he wouldn't feel the need to stoop so low to make [his] point. That really got me. I'd never heard of Boyd Haley before that day. I learned very soon that he was looked upon as quite a hero by a lot of people. That was what really drew my attention to this [group of scientists]; it was the mad child thing."

Seidel wasn't the only one to question Haley's expertise. When Haley appeared as an expert witness against vaccine makers, one judge refused to admit his testimony: "The court finds that Dr. Haley's report does not state an expert opinion that thimerosal causes autism," he wrote, "rather just that he has a theory about how such a thing could happen. At best, he expressed [his] 'strong belief.'" In January 2008, Haley's testimony was thrown out of court again, the judge ruling that "his lack of expertise in genetics, epidemiology, and child neurology make it impossible for him to supply the necessary factual basis to support his testimony." During that trial, Haley further damaged his credibility when, according to the judge, he "accused the Institute of Medicine committee of dishonesty and asserted that the CDC bureaucrats should be charged with criminal activity."

Undeterred by these failures in court, Haley designed his own brand of mercury chelators. During an online interview, a New Jersey physician who treated autistic children in a hyperbaric oxygen chamber in his office asked Haley, "Rumor has it you are developing a new chelator. Could you briefly fill us in?" "I have several," said Haley. "They're done. They're working great. We're doing biological testing now, making sure it's safe in animals. Then we'll put it into clinical trials. We're very excited about it." In 2001, Haley, convinced he had found a cause and a cure for autism, presented his work to an advisory panel of the Institute of Medicine (IOM) chaired by Marie McCormick. "In general, we did not find his work to be persuasive," she said. When asked about McCormick's assessment, Haley said, "I think there's a special place in hell for her."

Haley's entrance into the world of autism wasn't the first time he'd participated in a medical controversy. In the 1980s, a group of dentists claimed most facial pains were caused by cavities deep in the jaw bone; the disease was called neuralgia-induced cavitational osteonecrosis (NICO). NICO's principal promoter, a dentist named J. E. Bouquot, claimed that NICO caused all kinds of pains, even those located far from the mouth. Bouquot proposed that NICO could be treated by drilling into the jaw, scraping the bone, and rinsing the wound with antibiotics or, if that didn't work, with colloidal silver and vitamin C. With his new disease in hand, Bouquot founded Head and Neck Diagnostics of America in Morgantown, West Virginia—a lucrative mail-in biopsy service. NICO was an elusive diagnosis. Cavitations couldn't be found on traditional X-rays, so the diagnosis could be made only by biopsy. And the only biopsies found to be abnormal were those evaluated by Dr. Bouquot; other pathologists reviewing the same specimens couldn't find a problem.

NICO provided a financial windfall for the dentists who promoted it. Boyd Haley also participated, offering an "oral toxicity" test to detect NICO, a disease that didn't exist.

· · · ·

NEXT, SEIDEL TURNED HER ATTENTION TO THE GEIERS.

Mark and David Geier had been among the first to publish
papers claiming that thimerosal caused autism and that chelation therapy and Lupron helped. They testified in state legislatures
and in state courts, and they spoke at autism rallies, marches, and
conferences. They were the very centerpiece of the movement,
portrayed by the media as courageous men willing to take on
rich and powerful pharmaceutical companies, public health
agencies, and their own medical profession. But Mark and David Geier—like Andrew Wakefield before them—were not exactly what they appeared to be.

In 1983, after joining the faculty at Johns Hopkins, Mark
Geier launched Genetic Consultants, one of the first private companies licensed to perform in vitro fertilization. Geier aimed to please,
saying not only that he could provide a baby, but also, depending
on the parents' wishes, that he could provide a boy or a girl. He
was much better at making girls. "Out of twenty-five pregnancies,
twenty-four have been girls," said Geier. "[Now] I only do girls. I
used to do boys, too, but the success rate on them was only
seventy-five to eighty percent." But selective breeding didn't sit
well with the American public, so Geier moved into the expert
witness business on behalf of plaintiffs suing vaccine makers. His
new career began in December 1987, when he testified against
Connaught Laboratories, a pharmaceutical company based in Toronto and the maker of the diphtheria-pertussis-tetanus (DPT)
vaccine. Geier testified that the pertussis (or whooping cough)
part of the DPT vaccine "is one of the most feared poisons in the
medical sphere; [it] certainly can cause brain damage, death,
and many other things." Geier also believed that vaccines caused
a variety of chronic diseases; with his son David—president of
MedCon, a medical consulting company for lawyers and plaintiffs

interested in suing pharmaceutical companies—he later founded the Institute for Chronic Illnesses.

To investigate vaccines, Mark Geier converted the basement of his home in suburban Maryland into a personal scientific laboratory. While working on an article for the *New York Times*, reporters Gardiner Harris and Anahad O'Connor paid him a visit, later describing what they had seen: "Past the kitchen and down the stairs is a room with cast-off, unplugged laboratory equipment, wall-to-wall carpeting and faux wood paneling that Dr. Geier calls 'a world class laboratory—every bit as good as anything at NIH.'"

Mark (right) *and David Geier in their basement laboratory (courtesy of Marty Katz).*

By 2005, Mark Geier had testified in more than ninety trials against vaccine makers, but some judges considered him no more of an expert than Boyd Haley. In 1991, one judge stated, "Counsel and petitioners are forewarned of the court's position on Dr. Geier, or other doctors with little relevant experience, as an expert witness. In addition, the court admonishes Dr. Geier to reconsider his role, from an ethical and moral standpoint, as a witness. Several petitioners have lost cases solely on the little weight given to Dr. Geier's opinion." In 1992, a federal judge ruled, "I cannot give his opinion any credence." In 1993, a judge found that Mark Geier's testimony bordered on fraud: "[His] affidavit was seriously intellectually dishonest." Later that same year, yet another judge stated, "Dr. Geier's testimony is not reliable or grounded in scientific methodology and procedure. His testimony is merely subjective belief and unsupported speculation." And in 2003, a judge referred to Mark Geier as "a professional witness in areas for which he has no training, expertise, and experience."

While serving as a plaintiffs' witness in cases against the whooping cough vaccine, Mark Geier had seen an opportunity with autism. Richard Barr, the British personal-injury lawyer who had pursued the case against the MMR vaccine in England, paid Geier $14,000 to testify against vaccine makers. In 2003, a few years after he had received money from Barr, Geier published a paper claiming that the MMR vaccine caused autism. Geier said autism could be prevented if people used a different version of the vaccine, one that wasn't available: "In order to alleviate many of the difficulties encountered with the MMR vaccine, we suggest that a killed MMR vaccine should be made available as it may reduce the number and severity of adverse reactions following MMR vaccine." Geier failed to mention that a killed measles vaccine, sold in the United States between 1963 and 1967, had been withdrawn because it was found to be unsafe.

As study after study showed MMR didn't cause autism, Mark and David Geier went with the flow, now claiming that thimerosal was the problem. (The MMR vaccine never contained thimerosal.) The study that brought them immediate attention from the autism community was titled "Thimerosal in Childhood Vaccines, Neurodevelopmental Disorders, and Heart Disease in the United States." The Geiers had reviewed data from the Vaccine Adverse Events Reporting System (VAERS) and claimed that thimerosal in vaccines had caused autism, speech disorders, and heart attacks. But, as noted by the American Academy of Pediatrics (AAP), methodological flaws rendered the study worthless. First, the Geiers referred to VAERS as a mandatory reporting system for problems following vaccines. But VAERS is a passive system; people who believe that a vaccine might have caused a problem are encouraged, not mandated, to fill out a form. Because reporting is at best haphazard, most researchers (other than the Geiers) don't use VAERS to prove a vaccine has caused harm. Second, the Geiers claimed that thimerosal in vaccines put children at increased risk for heart attacks. But heart attacks are extremely rare in young children and the term *heart arrest*, although often used on death certificates, is usually unrelated to the disease responsible for the death. (Heart arrests are part of all deaths. When people die, their heart stops beating.) Third, the Geiers hadn't described any of the statistical methods they had used, a requirement for any legitimate medical or scientific paper. Finally, the Geiers claimed children were receiving more thimerosal in vaccines than ever before, when in fact thimerosal had been taken out of most childhood vaccines two years earlier.

The AAP didn't stand alone in its criticisms. The Geiers' article was criticized by the IOM, the CDC, and several academic institutions. Later, in a study published in *Pediatrics*, public health officials were disappointed to learn that reports of autism to VAERS weren't coming from parents, doctors, nurses, or nurse practitioners; they were coming from personal-injury lawyers. Mark Geier had used VAERS data to "prove" that thimerosal caused autism, and personal-injury lawyers pointed to those data as evidence that their clients had been harmed. For the lawyers, VAERS reports hadn't been a self-fulfilling prophecy; they'd been a self-generated prophecy. "Lawyers are manipulating this system to show increases that are based on litigation, not health research," said Michael Goodman, the *Pediatrics* study's lead author.

The medical journal that had published many of the Geiers' papers, the *Journal of American Physicians and Surgeons*, and the group behind it, the Association of American Physicians and Surgeons, were also illusory. Directed by Jane Orient, the association is run out of a small office in a strip mall in Tucson, Arizona. On October 14, 1999, Orient, appearing on ABC's *Nightline* with Ted Koppel, likened vaccines to scientific experiments in Nazi Germany.

No one discredited the Geiers more than a group of researchers headed by Sarah Parker and Jim Todd, from the Children's Hospital in Denver, and Ben Schwartz and Larry Pickering, from the CDC. In 2004, in a paper published in *Pediatrics*, these investigators graded the four epidemiological studies that had exonerated thimerosal as a cause of autism (written by Thomas Verstraeten, Anders Hviid, Jon Heron, and Nick Andrews) and the three studies that had claimed thimerosal caused autism (all written by Mark and David Geier). "We outlined eight epidemiological criteria that should be fulfilled," recalled Larry Pickering, the study's senior author. The Andrews study was considered to have adequately addressed seven of the eight criteria; the Heron paper, five of eight; the Hviid and Verstraeten papers, six of eight; and the three Geiers' papers, zero of eight. The authors concluded that the Geiers' papers were "of poor quality and cannot be interpreted." "None of the studies were perfect," recalled Pickering. "None of [them] fulfilled all eight of the epidemiological criteria, which is not surprising. But the Geiers' studies fulfilled none of them. They were very low quality studies."

A few months after publishing their paper, Sarah Parker, Ben Schwartz, Jim Todd, and Larry Pickering each received a certified letter in the mail. "We were being sued," recalled Pickering. Even though several judges had already thrown Mark Geier's testimony out of court, the lawsuit claimed the authors had defamed him, describing the Geiers as men who "dedicated substantial time and resources to serving as expert witnesses [and were] dependent upon the compensation they receive as a significant source of income." The Geiers were seeking more than $1 million in damages. Pickering remembered what happened next: "We had to deal with lawyers from the CDC and the University of Colorado and the American Academy of Pediatrics, which took a lot of time and effort and money." Eventually, the lawsuit was dropped. "[Our paper] was just considered part of the normal scientific discourse," recalled Parker. Pickering was angry that the Geiers wanted to participate in the scientific process only to the point of being criticized. "It was upsetting not so much for the lawsuit, because I knew it was frivolous," said Pickering. "But in medicine and science we have a process. If you don't like the results of a study you write a letter to the editor, [which] is then answered. It's interesting to me that the Geiers wanted to have it both ways. They wanted to publish their articles in what they considered to be scientific journals, [but] when they were questioned or criticized, they abandoned the scientific process and called their lawyers."

In 2004, the Geiers began to promote Lupron as a drug to treat autism. Lupron is approved by the FDA to treat prostate cancer, endometriosis (a disease that causes pain and abnormal menstruation), and precocious puberty (when children enter puberty much earlier than normal). The following exchange took place in a Washington, D.C. court on November 12, 2004, months after Mark Geier had already treated several children with Lupron. Geier, who was serving as an expert witness in a case against a vaccine maker, was being questioned by a defense attorney:

Q: Have you ever been involved in the diagnosis of
      autistic children?

A: No, not directly.

Q: Have you ever been involved in the care and treat-
      ment of autistic children?

A: Not before this week.

Q: Do you have any formal training in the field of
      toxicology?

A: No.

Q: Have you ever been involved in the care or treatment
      of victims of mercury poisoning?

A: Not until recently.

Q: And what is the recent episode you're alluding to?

A: We presented a new idea on how to treat autism and
      how to treat mercury poisoning, because these kids
      aren't autistic, they're mercury poisoned. Although
      I'm not happy with trying it on children without
      further research, these people are desperate and there
      have been some remarkable responses.

At the time, eight states also used Lupron to chemically cas-
trate sex offenders, reasoning that a decrease in testosterone
production would decrease sexual drive. (The Texas Council on
Sex Offender Treatment recommends Lupron for only the most
predatory and violent offenders.) But Lupron isn't licensed by
the FDA to treat autism. By using Lupron on autistic children
without sufficient training in pediatrics, endocrinology, or pharmacology, Mark Geier had crossed an important line.

In a thirteen-part series on her Web site, Kathleen Seidel took
a closer look at the Geiers and their claims. She described a
scene at a conference in 2005 sponsored by the parent advocacy
group Autism One, in which David Geier pointed to a slide
showing how testosterone could bind to mercury. "We discovered that testosterone and mercury complexed in the body," said
Geier. "You're looking at your testosterone here—that's these
folded sheety things. They form long strands, like little threads.

And [mercury] binds one string of testosterone to another." David Geier explained that it was hard to rid mercury from the body using chelation therapy alone because some of the mercury was embedded in testosterone. The only way to effectively get rid of mercury was to use chelation and Lupron simultaneously. Mark and David Geier were correct that previous research had shown that testosterone could bind to mercury. But they neglected to mention that binding occurred only in a test tube at very high temperatures and with the use of benzene, a toxic chemical not found in the body. The Geiers' Lupron therapy was based on a condition that doesn't happen in nature.

Seidel uncovered other unsubstantiated claims by the Geiers. Although Lupron is recommended by the FDA for treatment of precocious puberty, there is no evidence supporting the Geiers' claim that autistic children have abnormally high levels of testosterone or suffer precocious puberty more than other children. Seidel reasoned that using a drug to treat autistic children who didn't have precocious puberty was probably unethical and possibly dangerous. But the Geiers were reassuring. Speaking to parents at an autism conference, they said, "The first thing people think is, well, the side effect of the drug we use, which is called Lupron, is really bad. And it turns out it isn't." But Lupron can cause hives, difficulty breathing or swallowing, numbness, tingling, weakness, painful or difficult urination, blood in the urine, bone pain, testicular pain, and osteoporosis. This last problem isn't trivial; osteoporosis is also worsened by drugs that alter mineral metabolism, like chelation, and by diets free of dairy products. Further, the company that makes Lupron has warned: "No clinical studies have been completed to assess the *full reversibility of fertility suppression.*" This meant that Lupron, in addition to slowing normal sexual development, might cause autistic children to become sterile permanently. The company warned that Lupron should not be given to boys older than twelve years of age—a warning the Geiers didn't heed.

Seidel dug further. She wanted to find exactly who was supervising the Geiers' Lupron trials and how such a study could be permitted by a federally mandated Investigational Review Board (IRB). "In the abstract [of one of the Geiers' papers] it said that the Investigational Review Board for the Institute for Chronic Illnesses approved the present study," recalled Seidel. "And I thought who the hell would ever give ethical approval to [this] research? How can anyone conceivably give informed consent if they've been [told] that testosterone bonds with mercury in the body and that you can help an autistic person by [getting] the mercury out? These were disabled children, no less. And what the hell is the Institute for Chronic Illnesses? I wanted to find out who was on the board. So I filed an FOIA [Freedom of Information Act] request with the Office of Human Research Protections. I remember when I went down to the mailbox and there it was. I picked it up and my heart was pounding."

What Seidel found out about the IRB wasn't very reassuring. First, she discovered that it wasn't officially formed until fifteen months *after* the Geiers' study had started. Then she found the identities of the seven board members. The first two listed were Mark and David Geier, who clearly shouldn't have been judging the ethics of their own study. The third member, Lisa Sykes, was a Methodist minister and anti-thimerosal activist who had touted the wonders of Lupron at autism conferences and had volunteered her son to receive it. Sykes was in the midst of a $20 million lawsuit against GlaxoSmithKline, Wyeth, and Bayer Pharmaceuticals. The fourth member was Kelly Kerns, a dental hygienist, anti-thimerosal activist, and mother of three autistic children. Kerns had already sued the government for damages. The fifth member was John Young, an obstetrician and gynecologist. Young was Mark Geier's business partner in Genetic Consultants of Maryland and Genetic Consultants of Virginia. The sixth member was Anne Geier, Mark Geier's wife and David Geier's mother. The seventh member was Clifford Shoemaker, a vaccine-injury lawyer who had often used Mark Geier as an expert witness. Shoemaker had posted the Geiers' first publication on his Web site, praising their dramatic new therapy: "Unlike those in the government who want to bury their heads in the sand with comments like, 'We don't know what caused the autism epidemic, but we're sure it wasn't the mercury in vaccines,' Dr. Geier is spending his time trying to find ways to treat children with autism. Here is his hypothesis, and so far it's batting a thousand percent. Watch for more exciting developments."

This, concluded Seidel, was not an independent review board.

Seidel found something else strange about the Geiers' papers. She wondered why they referred to their revolutionary new therapy by its trade name (Lupron) rather than its generic name (leuprolide). Typically, because medications are often made by more than one company, researchers use generic names in their publications. Leuprolide, for example, is made by three companies: Sanofi Aventis, which calls it Eligard; Bayer Pharmaceuticals, which calls it Viadur; and TAP Pharmaceuticals, which calls it Lupron. Puzzled, Seidel wrote a letter to the medical journal that was about to publish one of the Geiers' Lupron papers. "Dr. and Mr. Geier identify leuprolide as LUPRON®—all capital letters, registered trademark on prominent display—no fewer than fourteen times in their paper, and use the generic name only twice," wrote Seidel. "Review of a half-dozen academic journal articles on leuprolide yields no instances in which a brand name is displayed so frequently and unnecessarily, contrary to conventions for the use of trade versus generic names." Then Seidel discovered why the Geiers might have chosen to specifically promote Lupron. She found two patents on Lupron as a treatment for autism submitted by the Geiers; one had been filed in September 2004, the other in September 2005. "I read these patent applications and they're deadly boring, incredibly boring," recalled Seidel. "The first one was obviously prepared by the Geiers themselves. The second one they had the help of a patent attorney in Chicago. I didn't recognize the significance of Chicago until later on, when I found that that was the IP [intellectual property] firm for TAP Pharmaceuticals. TAP's lawyers had obviously prepared the application. They were not a co-applicant on the U.S. patent application but later had published an international patent application that had TAP's name on it."

The Geiers, who had entered into a licensing partnership with TAP, hoped that Lupron would eventually be approved by the FDA for the treatment of autistic children; if it were, it could bring millions of dollars in royalties. Seidel wrote a letter to TAP Pharmaceuticals explaining how its product, which was licensed by the FDA for the treatment of precocious puberty and prostate cancer, was instead being offered by the Geiers as a treatment for autism. TAP's staff attorney and director of marketing told Seidel that the company had no control over the off-label use of their drug.

TAP Pharmaceuticals—which stands for Takeda Abbott Pharmaceuticals, a joint venture between Takeda Chemical Company in Osaka, Japan, and Abbott Laboratories in Abbott Park, Illinois—wasn't a stranger to controversy. Several years before, to compete with the rival drug Zoladex, a product of Astra Zeneca, TAP had given free samples of Lupron—a drug that cost about $20,000 per treatment course—and offered grants of $25,000 to any physician who agreed to stop using Zoladex. The Department of Justice saw this marketing practice for what it was: fraud. It charged fifteen employees and five physicians for fraudulently promoting Lupron. On October 3, 2001, TAP agreed to pay almost $900 million to settle the government's criminal complaints. It was the largest settlement for health care fraud in U.S. history.

Determined to expose the Geiers, Seidel wrote to medical professional societies, ethics boards, members of the National Immunization Program at the CDC, and the vice-president of marketing at TAP Pharmaceuticals—all to no avail. Later, she lamented a culture that permits autistic children to be harmed by people like Mark and David Geier. "It is difficult to comprehend how a doctor with no advanced training in autism, toxicology, pharmacology, or endocrinology, assisted by a young man with no higher professional credential than a B.A. in biology, could persuade loving parents to subject their growing autistic children to the powerful hormonal inhibitor, Lupron," she wrote. "It is less difficult to conceive, however, when one considers the 'autism-biomed' subculture in which the Geiers and their associates have conducted their publicity and recruitment efforts. The subculture consists of parents who attribute their children's autism to vaccine injury; doctors and manufacturers who market products and services to these families; lawyers and political activists who encourage parents to seek legal redress for their children's disabilities; and an insular, mutually enforcing, minority of academics, many of whom have received substantial research funding from organizations advocating for plaintiffs' interests, and some of whom are serving as expert witnesses."

. . . .

SEIDEL WAS ALSO APPALLED BY THE GEIERS' AND J. B. HANDLEY'S constant promotion of chelation therapy as a cure for autism. In 2000, only a handful of children were chelated; by 2005, the number had purportedly climbed to more than 10,000 a year. On her Web site, Seidel pleaded with Handley to stop promoting a "therapy" that had never been shown to work and was potentially dangerous. Handley wrote back: "I have no respect for your movement," he said. "We are spending our time constructively engaging doctors to help our babies. If you don't like what we have to say, stop listening. We will bring the full resources of myself and Generation Rescue to stop this. We will sue you for libel and will go after your home and assets. My lawyers live to investigate and sue people like you. This will be your only warning."

When Lyn Redwood and Sallie Bernard proposed that mercury in vaccines caused autism, no evidence existed to support or refute their claim; it was just speculation. But during the next seven years, scientists on several continents took this hypothesis seriously enough to examine it and found it was wrong. More important, the notion that chelation therapy can heal the brains of children who had actually suffered mercury poisoning is false. Once a brain cell has been damaged by a heavy metal like mercury, it is permanently damaged. The IOM, during its review of thimerosal, stated: "Chelation therapy has not been established to improve [kidney] or [brain] symptoms of chronic mercury toxicity and has had no effect on cognitive function when used for excretion of another heavy metal, lead. Because [chelation therapy] is unlikely to remove mercury from the brain, it is useful only immediately after exposure, before damage has occurred. Moreover, chelation therapy is not without risks; some chelation therapies might cause the release of mercury from [other parts of the body], thus leading to *increased* exposure of the [brain] to mercury." So, it's not only that chelation therapy hasn't been shown to work; it's also that it didn't make sense that it would work. J. B. Handley had founded an organization dedicated to proselytizing the wonders of chelation therapy, recruiting scores of Rescue Angels to spread the word. And many parents swear by the wonders of chelating autistic children, claiming dramatic improvement within days. But because a cell damaged by heavy metal doesn't recover—much less within a few days—this is simply not possible.

Then the unthinkable happened. On the morning of August 23, 2005, Marwa Nadama brought her five-year-old autistic son, Tariq, to the Advanced Integrative Medicine Center in Portersville, Pennsylvania, about thirty-five miles northwest of Pittsburgh. Marwa, who was born in Nigeria but had moved to London in 1995, had been frustrated by the unwillingness of British doctors to chelate her son. So she called a DAN doctor who recommended that Marwa take her son to Dr. Roy Kerry, a sixty-eight-year-old ear, nose, and throat specialist. At around ten o'clock, Theresa Bicker, a medical assistant working with Dr. Kerry, gently took Tariq's arm, rolled up his sleeve, cleaned an area of skin with alcohol, inserted a needle attached to a syringe, and injected ethylenediaminetetraacetic acid (EDTA), a chelating medicine, into his bloodstream. Because Tariq had struggled during the procedure, Theresa had requested that another doctor working in Dr. Kerry's office restrain him. During the procedure Marwa noticed that something was wrong with her son. She called out to the doctor, who took Tariq's blood pressure. After Tariq went limp, Bicker immediately called 9-1-1. The ambulance arrived minutes later and rushed Tariq to Butler Memorial Hospital, where he was pronounced dead of a heart attack. The chief of

forensic pathology later ruled that Tariq had died from dangerously low levels of calcium in the bloodstream due to administration of EDTA. EDTA doesn't bind only to mercury; it also binds
to calcium, which is important in conducting electrical impulses
within the heart. When Tariq's calcium level dropped precipitously, his heart stopped beating. Tariq Nadama was the third
person to die from EDTA therapy since 2003. But J. B. Handley's
Rescue Angels weren't about to quit. "We're not stopping," said
one of them, Marla Green.

In September 2006, the Department of State, which licenses
physicians in Pennsylvania, filed six disciplinary charges against
Roy Kerry for his role in the death of Tariq Nadama. One year
later, on August 22, 2007, the district attorney's office charged
Kerry with involuntary manslaughter and reckless endanger-
ment; he was told to turn himself in immediately or risk arrest.
The lawyer for the Nadamas, John Gismondi, noted the unusual
nature of the charges: "Most medical situations don't involve
criminal charges. But [in this case] I think criminal charges are
warranted." The case against Roy Kerry was later dropped.

• • • •

IN ADDITION TO BOYD HALEY AND MARK AND DAVID GEIER,
two other scientists had stepped forward to support the notion
that mercury caused autism: Richard Deth and Mady Hornig.

Richard Deth was the biochemist at Northeastern University
who had proposed that the response of laboratory cells to thimerosal provided a biological basis for autism. Deth knew autism
was a problem with nerve cells from the brain, not nerve cells
found in muscles or organs. He also knew that autism wasn't
associated with cancer of the brain. And he knew that autism
affected cells that had the normal number of chromosomes
(strands of genetic material that contain the blueprint for cells
to function and reproduce). Nonetheless, Deth chose to study
cancerous nerve cells from outside the brain that had an abnormal number of chromosomes. Deth treated these cells with alcohol, lead, thimerosal, and aluminum. He found that all of these

substances affected an important metabolic pathway. But when he published a paper on his findings, Deth chose the title "Activation of Methionine Synthase by Insulin-Like Growth Factor-1 and Dopamine: A Target for Neurodevelopmental Toxins and Thimerosal." Given that almost everything he tested affected the metabolic pathway he was studying, why did Deth choose to single out thimerosal in the title? Why didn't he acknowledge that thimerosal, alcohol, aluminum, and lead caused the same reaction? His choice might have been influenced by his source of funding. On July 9, 2003, Richard Deth received $60,000 from Safe Minds, some of whose members were in the midst of lawsuits against the federal government and pharmaceutical companies claiming that vaccines had caused harm. Not only had Deth been paid by Safe Minds, but also he, like Mark and David Geier and Boyd Haley, had been retained as an expert witness. If Deth could implicate thimerosal as a cause of autism, he could help those who were paying for his research as well as his testimony.

Mady Hornig followed the path of Richard Deth. In 2003, she also published a paper showing that a particular breed of mice, highly susceptible to autoimmune diseases, developed autistic symptoms after they were injected with thimerosal. Hornig's choice to use mice with severe abnormalities of the immune systems was a poor one. Children with autism have never been shown to have brain abnormalities consistent with an autoimmune disease (as is seen, for example, in multiple sclerosis). Hornig presented her data to researchers at the IOM, who later stated the obvious: "The relevance of rodent models is difficult to assess because the rodent 'clinical' endpoints may not reflect the human ones [and] may bear no relationship to pathogenesis of human disease." Two years later, researchers at the University of California at Davis and the NIH, using the same strain of mice, failed to find what Hornig had found. On November 24, 2003, Mady Hornig received $35,000 from Safe Minds to determine whether thimerosal was a cause of autism.

. . . .

LENDING THE CREDENCE AND POPULARITY OF HIS FAMILY'S NAME, Robert F. Kennedy Jr. had also been a zealous advocate for parents convinced that vaccines had poisoned their children. When Kennedy had appeared on *Imus in the Morning* in June 2005, he had explained: "I got into this because I'm an attorney for the Natural Resource Defense Council and I'm president of the Waterkeeper Alliance. I've been an environmental advocate for twenty-one years. In the course of that, I got approached by these autism mothers, who made the point to me that the levels of environmental mercury were dwarfed by [that which] children were getting from vaccines." In his explanation to Imus, Kennedy had omitted a few facts about how he had become an activist.

In 1983, following a conviction for possession of illegal drugs, Kennedy was sentenced to two years' probation, periodic drug testing, mandatory supervision by Narcotics Anonymous, and 800 hours of community service. He satisfied his community service by working for the Hudson River Foundation, now called the Hudson Riverkeepers. Later, Kennedy became its chief prosecuting attorney.

Walter Olson has followed the career of Robert F. Kennedy Jr. Olson is the author of *The Rule of Lawyers* and *The Litigation Explosion*. Called "an intellectual guru of tort reform" by the *Washington Post*, Olson has helped to shape the debate on tort reform, and his publications have been cited in Supreme Court opinions. "[Kennedy] is a number of different things," said Olson. "He is a professor at Pace University Law School. He is the chief prosecuting attorney for the Hudson Riverkeepers. And he is also an attorney with one of the largest and best mass tort law firms in the country." It is this last association that has most intrigued Olson.

In 2000, Kennedy joined a group of trial lawyers to sue pork producers in the South and Midwest. During the litigation, the Associated Press ran an article that quoted Kennedy as well as a lawyer named Mike Papantonio. Although Kennedy and Papantonio were described as representing two different organizations, their appearance in the same article wasn't a coincidence. "Michael Papantonio is a flamboyant, very well known personal-injury lawyer who has been involved in a lot of mass tort cases," says Olson. "He is the second named partner in the law firm of Levin Papantonio, one of the best known mass tort firms. They were involved in tobacco litigation. They were involved in asbestos. They have eighteen different product-liability areas that are important enough for them to list [on their Web site]. They are a very rich, very successful firm. And they're not a firm that incidentally does plaintiffs' work. They are very close to the glowing heart of the product-liability industry. A firm of that sort, even if it doesn't list vaccine litigation as one of its major activities, is well aware of the large amounts that could be made if it cracks open. The firm doesn't list what Kennedy's financial arrangements are, but that there are financial arrangements is certainly implied by the fact that he is 'of counsel' to the firm."

Olson noticed that Kennedy's association with the Levin Papantonio law firm wasn't mentioned during the thimerosal-autism debate. "It seems to me that editors at places like *Rolling Stone*, if they were doing their jobs, would probably want to inform readers that Kennedy was a participant in a law firm that regularly sued drug companies and could potentially open up a new lucrative area of law," he said. "I think that it's much more likely that they think, 'Gee, we get to print an article by Robert F. Kennedy Jr. Let's take whatever byline he gives us to explain his current status in society.'"

Kennedy still lends his celebrity to the mercury-autism controversy, occasionally writing articles for the *Huffington Post*, and he continues to host a radio show on Air America called *Ring of Fire*. His co-host is Michael Papantonio.

· · · ·

IN 2003, WHILE WORKING ON *EVIDENCE OF HARM*, THE BOOK THAT became the centerpiece of the anti-thimerosal crusade, David Kirby went to the CDC to interview public health officials. First, he had to gain permission from Curtis Allen, a senior press officer at the CDC. Allen denied Kirby access. In the introduction to *Evidence of Harm*, Kirby was bitter about the experience: "Many of the public health officials who discount the thimerosal theory were unwilling to be interviewed for this book, or prohibited from speaking by their superiors," he wrote. Although Kirby later used Curtis Allen's prohibition as evidence of a conspiracy of silence, Allen stands by his decision. "[Kirby] called me and said that he was a reporter for the *New York Times*," recalled Allen. "And he said that he wanted to interview Walt Orenstein and a couple of others. I had never heard of this reporter. I had never talked with him. Then he said something that I thought a legitimate reporter would never say. He made such a big deal out of being a *Times* reporter. 'I'm with the *New York Times*. I'm with the *New York Times*' he kept saying. You only have to say it once. I've got ears. I can hear. But he kept pounding the point home. So I called the *New York Times* and asked a couple of reporters that I know whether they had ever heard of David Kirby. And they said 'No.' So we did a Google and LexisNexis search, and the only articles that we could find showed that he was a freelancer [and occasional contributor to the *Times*] and not on the staff at any newspaper. [Kirby, who owned and directed a public relations firm in New York City, had written several freelance articles about art, bars, and aircraft.] When Kirby called back, I said that the interviews were off because it appeared he had misrepresented himself. He immediately started threatening me, saying that he'd call the editorial board at the *New York Times*, and what did I think about that? I told him, 'Tell them to give me a call; they have my number.'"

Kirby had other shortcomings as a journalist. Although *Evidence of Harm* was embraced as a manifesto for those who believed thimerosal caused autism, David Kirby didn't portray himself as a flag-waver for their cause. "I am a journalist, not a politician," he said. "I don't want to take a position one way or another." Arthur Allen, the journalist who had written "The Not-So-Crackpot Autism Theory" for the *New York Times*, recounted how he had first learned that Kirby was writing his book. "I remember talking to Lyn Redwood when the *Times* piece came out," said Allen. "And she told me she was working on a book about thimerosal. [Months later], I asked her how her book was coming, and she said she had found a writer to do it." That writer was David Kirby.

When Arthur Allen met David Kirby in 2004, several epidemiological studies had already been published showing that thimerosal didn't cause autism. Allen, convinced by the data, no longer believed the theory he had first posited in the *New York Times* in 2002. He was surprised that David Kirby, whose book wasn't published until April 2005, still did. "In 2004 I met Kirby at an Institute of Medicine meeting," recalled Allen. "And I asked him 'What's the book? How are you going to write a book about this?' Because my view by then was there was quite a bit of evidence that shot [the thimerosal-causes-autism theory] down. And he said, 'I'm reporting the controversy.' I was just really skeptical about that. As a journalist, I have very little respect for him. He saw an opportunity and he took it. In my opinion, that's the lowest form of journalism." Allen doubts Kirby believes it. "I spent a certain amount of time with him when we were doing the debate," recalled Allen, who had debated Kirby in San Diego. "He kept saying that the data coming in from California [were] really strengthening [my] argument. As I understand it, most of Kirby's work has been in public relations. I feel that what he's doing is public relations—flacking for a really irresponsible cause."

Kirby's advocacy led him to speak at rallies, conferences, and marches on behalf of Safe Minds. In his book, Kirby had been circumspect about his decision to tell one side of the story. "Perhaps this story will be told one day from the opposing view," he wrote, "that of the doctors, bureaucrats, and drug company representatives who claim nothing more than the laudable desire to save kids from the ravages of childhood diseases." But after his book was promoted by Safe Minds and other parent groups whose members were in the midst of litigation, Kirby became a standard-bearer and, like the groups for whom he fronted, a conspiracy theorist. In 2005, both Al Franken's Air America and NPR refused to interview him. "NPR and the Public Broad-casting System get a lot of money from drug companies," said Kirby. "And they need whatever money they can get. So they are not going to offend any advertiser—ever; whereas the major com-mercial networks have a little more leeway and play. They take more risks. The conservative press is anti-government, whereas the liberal press is so pro–public health—it is like the Centers for Disease Control can do no wrong. Doctors can do no wrong." Kirby was angry at the snub by NPR: "The national NPR has ignored this book, hung up on me, written me back and told me to take them off my mailing list," he said. "Never in fifteen years as a journalist have I ever been treated like this by any-body, except for the CDC."

In April 2007, after Don Imus was fired from MSNBC for a disparaging remark about the Rutgers women's basketball team, David Kirby came to his defense, seeing more evidence for con-spiracy. "Imus is gone, but not everyone is cheering," wrote Kirby. "Thousands of parents of autistic children around the country are reeling at the loss of the one true friend they had in the mainstream media. For them, the silencing of Imus could not have come at a worse time. And, of course, if anyone is happy to see Don's down-fall, they are also in the fine company of Eli Lilly, Merck, Glaxo-SmithKline, and other big Pharma firms who loathed Don Imus for suggesting that mercury in vaccines might be contributing to the growing crisis of childhood autism in the United States. Maybe mercury is linked to autism and maybe it is not. But until we find out, go ahead and buy that Lilly stock you've had your eye on. With Imus out of the picture, your investment is safer."

· · · ·

DURING THE CONGRESSIONAL HEARINGS IN APRIL 2000, CONGRESS-man Henry Waxman, in response to Dan Burton's contention that the MMR vaccine caused autism, had said: "Let us let the scientists explore where the real truth may be." Waxman be-lieved that the question of whether vaccines caused autism was a scientific one, best answered in a scientific venue. But many

Don Imus during an appearance on Al Sharpton's radio show to apologize for derogatory remarks against the Rutgers women's basketball team, April 9, 2007 (courtesy of Getty Images).

people don't believe that. They believe that any issue should have its day in court—that judges and juries should be the final arbiters of scientific truths. In 2001, Lyn and William Redwood, on behalf of their son Will, filed a lawsuit under the Court of Federal Claims in Atlanta, Georgia; they wanted the federal government and vaccine makers to pay for their son's care. It was just the beginning. Soon parents of thousands of autistic children followed the Redwoods' lead. During his hearing on September 8, 2004, Dan Burton warned pharmaceutical companies of the growing litigation: "When, and I'm not saying if, but when it's proven that the mercury in vaccines has been a major contributing factor to those damaged kids, then there's going to be a tremendous amount of liability exposure for these pharmaceutical companies and then they're going to be out there all by themselves."

On April 4, 2005, David Kirby, also aware of the lawsuits, made a plea on *Imus in the Morning*. Because neither parents nor states could afford the therapies, reasoned Kirby, pharmaceutical companies should pay for them: "We need to figure out how we're going to compensate these families; how we're going to take care of these children; how we're going to remove the burden from states, because right now they're footing the bill for everything. I don't want to see the drug companies go out of business. I don't think anyone wants that. [But] we are looking at trillions and trillions of dollars of care for these people." One trial lawyer likened pending lawsuits against vaccines to tobacco litigation. "If you think that cigarette companies were hit hard," he said, "wait 'till these vaccine cases go to the juries. The jurors who awarded millions of dollars to the plaintiffs had limited sympathy for them, since the risks of emphysema and lung cancer were well known. But children didn't choose to be vaccinated—and their lives and their family's lives have been ruined."

David Kirby, Dan Burton, Robert F. Kennedy Jr., personal-injury lawyers, and parents were angry and frustrated. They wanted someone to pay for the vitamin injections, blood tests, anti-fungals, anti-virals, mineral supplements, cranial manipulation, special diets, sonar depuration, hyperbaric oxygen, Lupron, and chelation. So they turned to the two groups that had those resources: pharmaceutical companies and the federal government. It was time for them to pay for the damage they had caused.

On June 11, 2007, the vaccine-autism theory had its first big day in court.

CHAPTER 8

Science in Court

And there went out a champion from the camp
of the Philistines, named Goliath of Gath,
whose height was six cubits and a span.

—1 SAMUEL 17:4

In the summer of 2007, parents of children with autism took their case to court. Called the Omnibus Autism Proceeding, it was an unusual lawsuit. Parents weren't suing the company that made thimerosal (Eli Lilly) or the company that made MMR (Merck) or the companies that made vaccines containing thimerosal (Merck, GlaxoSmithKline, Wyeth, Sanofi Pasteur, and Novartis). They were suing the federal government in a federal court. This wasn't their preference. They would much rather have argued their cases in state courts in front of juries. In federal court they would have to convince a panel of three judges. But they had no choice; no one can sue a vaccine maker without first going through this unusual court.

• • • •

IN 1986, FOLLOWING A SERIES OF LAWSUITS THAT THREATENED to end vaccine manufacture for American children, Congress passed the National Childhood Vaccine Injury Act. Included in the act was the Vaccine Injury Compensation Program. If parents felt their children had been harmed by vaccines, they sued the federal government for compensation, making their arguments in front of federally appointed judges. As a consequence, the number of lawsuits brought against vaccine makers declined dramatically.

At the heart of the Vaccine Injury Compensation Program sits a list of known vaccine injuries. If studies have clearly shown that a vaccine had caused harm, children are compensated. For example, Albert Sabin's live weakened polio vaccine—the one that had been dropped onto sugar cubes and given to children in the United States between 1962 and 1998—was a rare cause of paralysis. Every year, between six and eight American children were paralyzed by this vaccine. Under the federal compensation program, children harmed by Sabin's vaccine were compensated quickly. They didn't have to take their case to a state court, where the process is slower and more expensive. On the other hand, claims not supported by scientific evidence, such as paralysis caused by chickenpox vaccine or diabetes caused by Hib vaccine, haven't been compensated.

The Vaccine Injury Compensation Program is designed so that a decision can be reached within 240 days (so it's quick); the average compensation is about $900,000 (so it's generous); and the program is based, for the most part, on a preponderance of sound scientific evidence (so it's fair). But the federal program doesn't completely protect vaccine makers. If petitioners are unhappy with the judge's ruling in federal court, they can always take their cases to juries in state courts. This possibility scares pharmaceutical companies, and for good reason. Juries have historically been poor judges of scientific and medical truths. One notable example was the case against Bendectin, a drug that treated morning sickness in pregnant women. Bendectin was driven off the market in the 1980s by lawsuits claiming that it caused severe birth defects. At the time of these jury verdicts, twenty-seven separate studies had shown the incidence of birth defects was the same in women who did or didn't take the drug.

• • • •

VACCINE COURT WAS ORIGINALLY DESIGNED TO HANDLE ONE CASE
at a time. But between 1999 and 2007, more than 5,000 parents
filed claims that vaccines had caused their children's autism.
This was twice the number of claims filed for all other vaccine-
related injuries in the twenty years since the program had begun. Because of the number of claims and because the federal
judges knew that it would be impossible to hear each case individually, they recommended that autism claims be tried together,
like a class-action lawsuit. "There's never been another case like
this," said Kevin Conway, one of the lawyers for the petitioners.
With average individual awards of close to $1 million and thou-
sands of petitioners, it was possible that a ruling in favor of the
petitioners could exhaust the $2 billion available to compensate
claimants. Much was at stake.

• • • •

PETITIONERS CLAIMING VACCINES CAUSED AUTISM HAD THREE DIF-
ferent theories: MMR caused autism; thimerosal caused autism;
the combination of MMR and thimerosal caused autism. The
federal judges decided to try the last theory first. To represent
the theory that MMR and thimerosal together caused autism,
the petitioners chose Michelle Cedillo, a twelve-year-old girl
from Yuma, Arizona.

On June 11, 2007, George Hastings, the judge in charge of
the Cedillo trial, called the court to order. In the late 1980s,
Hastings, a graduate of the University of Michigan Law School,
had served as an assistant chief for the Tax Division of the U.S.
Department of Justice. Before that he had served in the Appeals
Division, and before that he had been the news editor for a television program called *Good Morning Michigan*. Sitting to Hastings's left was Denise Vowell, a recently retired chief trial judge
for the U.S. Army. Vowell was a member of the Women's Army
Corps and an officer in the Military Police. On Hastings's right
sat Patricia Campbell-Smith. During the 1990s, Campbell-Smith

had worked for Liskow and Lewis, a New Orleans law firm specializing in environmental law. The judges decided Hastings would be in charge of the first trial, and Vowell and Campbell-Smith would be in charge of the next two. All three judges had one thing in common: none had a professional background in science or medicine. Hastings was a tax lawyer, Vowell was a military lawyer, and Campbell-Smith was an environmental lawyer. Although scientists had already rendered a verdict on whether MMR or thimerosal caused autism, these three judges would be the final arbiters of scientific truth.

Federal Claims court, located across from the White House on Lafayette Square, seats 400 people. But despite requests by parent advocacy groups to hold the hearing in a much larger venue, only fifty showed up—mostly lawyers and journalists. After his opening remarks, Hastings described how the proceedings would work. "Today we are here for two purposes. One purpose is to hear the claim under the Vaccine Act of Michelle Cedillo. Michelle is a twelve-year-old girl who lives in Arizona and who has been diagnosed with autism and a number of other medical conditions. The first purpose of this hearing is to determine whether Michelle's autism and her other medical conditions were vaccine-caused.

"However, there is another equally important purpose to this hearing. That is, Michelle is one of nearly five thousand children diagnosed with autism or similar disorders to have filed claims under the Vaccine Act. These five thousand claims have been grouped together in a joint proceeding known as the Omnibus Autism Proceeding.

"The committee of attorneys who represent the petitioners in the Omnibus Autism Proceeding has designated Michelle's case as the first test case in that proceeding. Therefore, at this hearing today and over the next three weeks, we will hear not only about Michelle's own condition, but also extensive expert testimony concerning the petitioners' first general causation theory; that is, the general theory that MMR vaccines and thimerosal-containing vaccines can combine to cause autism."

Thomas Powers, from the Portland, Oregon, law firm of Williams, Love, O'Leary and Powers, was the first to speak on Michelle Cedillo's behalf. Powers's firm had successfully filed lawsuits against the makers of silicone breast implants, as well as against Fen-Phen, a weight loss product associated with heart problems, and the Dalkon Shield, an intrauterine birth control device found to cause severe infections and infertility. Class-action awards for the Dalkon Shield had totaled more than $2 billion; for breast implants, nearly $5 billion; and for Fen-Phen, $21 billion (and counting). These awards had made Williams, Love, O'Leary and Powers one of the richest, most powerful law firms in the United States.

Sylvia Chin-Caplan, from the Boston law firm of Conway, Chin-Caplan and Homer, also represented Michelle Cedillo. Chin-Caplan's firm specialized in claims before vaccine court, which allows lawyers to receive only 4 percent of awards. Trying cases before vaccine court isn't a very good way for personal-injury lawyers to make a lot of money. So unlike Thomas Powers and his partners, Chin-Caplan's firm was neither rich nor powerful, operating out of a modest three-story walk-up downtown.

After he made his opening remarks, George Hastings probably hoped that the lawyers would discuss Michelle's case in a collegial, friendly manner. It wasn't to be. Powers began by accusing the federal government and pharmaceutical companies both of hiding information and of collusion. "From day one," said Powers, "the [federal government] and industry have been on the same side of the table standing shoulder-to-shoulder doing everything they can to make sure that this climb toward proving causation is as long and as steep and as hard as it can possibly be." Powers was also upset that he couldn't bypass federal vaccine court, having failed to get autism cases in front of a jury in Oregon. "Way back in 2002 before the [Omnibus Autism Proceeding] was set up, some families had filed lawsuits in the civil justice system asking the courts and particularly asking juries to decide the issue of whether they had been injured by vaccines, suing the pharmaceutical industry and the vaccine manufacturers directly." Powers was a veteran of massive litigation against pharmaceutical companies; he knew where the money was. In Federal Claims court, awards would be much smaller. "As one would expect—and I totally expected it," continued Powers, "pharmaceutical industry lawyers were on the other table telling the federal judge to dismiss the case and send these children out of the courthouse. They shouldn't have a claim in front of a jury and they should instead come to the Vaccine Program. What I didn't expect is that the U.S. Government would stand literally, physically shoulder-to-shoulder with industry telling a U.S. District Court judge that these children ought to be tossed out of [civil] court and they ought to come here, taking the side of industry from day one."

Then Powers reprised a theme sounded by parent advocacy groups, Robert F. Kennedy Jr., and David Kirby: the government had hidden data that would have proved his case. "Something else that we've seen happen in the last five years in this program is a simple inability to get important, critical information and evidence," said Powers. "In the criminal justice system there's a process called discovery. It's available as a matter of right. If a party for litigation believes that somebody on the other side has relevant information, material information, they're entitled to simply ask for it and they get it. And if they don't get it, the judge tells the other side you've got to cough it up. [But] there's no right to discovery in this program. A lot of the evidence and a lot of the information on science and medicine are controlled by the federal government, but we can't get them. They have it. They're generating it. And we largely cannot get it." Powers was setting up a justification for an appeal. If his side didn't win this case, he was saying, it was because the government didn't supply the documents necessary to win it.

Vincent Matanoski represented the defense. A former captain and judge advocate general for the U.S. Navy Reserves and now a lawyer for the Department of Justice, Matanoski was angry at Powers's accusations of collusion and cover-up. Abandoning his original opening statement, he responded to Powers: "Mr. Powers tried to present the government in that federal case in Oregon [as] standing shoulder-to-shoulder with vaccine manufacturers. Well, I happen to be the government attorney who appeared in Oregon, and I remember very distinctly who I was standing shoulder-to-shoulder with. I was sitting next to Mr. Powers and his partner Mr. Williams at the table. And, in fact, my case, in very pertinent parts, stood opposite to several points that the vaccine manufacturers were making. [Mr. Powers] then complained about the frustrations the [petitioners] have had in discovery. In fact, the [petitioners] have received more data from the government in these proceedings than in all other vaccines cases combined over the twenty-year history of the program. He has received over 218,000 pages of government documents. He complains that the answers weren't in them. These were the documents he requested. These were the documents the [petitioners] sought."

Any semblance of collegiality between the government and the petitioners was shattered by the opening thrusts and parries of the lawyers. Now it was time to determine whether Michelle Cedillo's autism was caused by vaccines.

. . . .

ON THE FIRST DAY OF THE TRIAL, MICHELLE'S PARENTS ROLLED her into the hearing room in a wheelchair. Gardiner Harris of the *New York Times* described the scene: "She wore hearing protection similar to that worn by heavy machinery operators. She hit herself repeatedly and made loud grunting noises." During the previous year, Michelle's parents—Theresa, a homemaker, and Michael, a utility worker—had spent more than $18,000 taking care of their daughter. Everyone sitting in the courtroom on that first day was moved by the desperate nature of Michelle's condition and by the singular devotion of her parents to helping her.

Chin-Caplan began by describing the sad, isolated life of Michelle Cedillo: "Michelle was born on August 30, 1994. She weighed eight pounds, roughly, and her Apgars [a ten-point scale of a baby's health measured one and five minutes after birth] were nine and nine. In other words, she was perfectly healthy. [The day] after she was born, she received a hepatitis B immunization. It contained 12.5 micrograms of mercury. Her parents didn't know about [the danger of] it. The majority of the health profession didn't know about it.

"Michelle went for her regular doctor visits. This was the first child. This was the only child. They wanted this child very badly, and they were going to give her the very best medical care that she could ever have. They gave her all her immunizations because that was what was recommended. One month after she was born she went for hepatitis B immunization number two, another 12.5 micrograms of mercury. So we now have 25 micrograms of mercury, a cumulative dose in a child who is only one month old. By the age of seven months Michelle had received three DPT immunizations, three hepatitis B immunizations, and each DPT combined with [another vaccine] contained 25 micrograms of mercury. So you add up the math. During this period of time Michelle seemed to be okay. The pediatrician didn't think that there was anything wrong with her.

"In December 1995, Michelle went in for another immunization. She went for her MMR. One week later, Michelle developed a fever of 105. Her mother called the doctor and she was told there's a flu going around, a very bad flu. Keep her home. Nurse her, and she'll recover. The fever stayed up there; 105 on December 27; 105 on December 28; 105 on December 29; 105 on December 30 and yet she was told it's the flu. Finally, on December 31 it broke. And her mother thought. 'Thank God, it's finally gone.'"

Michelle's fever wasn't gone for long. On January 5, 1996, it returned, lasting two more days. Chin-Caplan continued: "After that, Michelle's family noticed that she wasn't speaking. She was totally silent. Before that she had been interacting with her parents, with her grandparents. She began interacting with her cousins. She was babbling. She was reaching for her toys. She was practically walking. She was sitting up by herself. [But then] she entered a little world of her own. She started engaging in repetitive behavior. The family would say 'Michelle, Michelle!' She ignored them like she had never heard them." Eighteen months after Michelle received MMR, in July 1997, doctors told Theresa and Michael Cedillo that their daughter had autism.

Michelle's troubles didn't end there. She began to have frequent bouts of diarrhea, vomiting, and stomach pain. Theresa Cedillo, frustrated that her daughter's symptoms were worsening, searched the Internet for answers. There she came upon DAN. She was excited to read about a DAN meeting in her area—a meeting where British gastroenterologist Andrew Wakefield would be speaking. "Theresa went to this conference," said Chin-Caplan, "and she stood in the back of the room and she listened to Dr. Wakefield talk. She stood in the room and she waited for him to finish speaking so that she could catch him and try to get his attention about helping her child." Eventually, Theresa brought Michelle to see Dr. Arthur Krigsman, who was now Andrew Wakefield's partner at Thoughtful House in Austin. In January 2002, Michelle had an intestinal biopsy that, according to Unigenetics (John O'Leary's laboratory in Ireland), contained measles virus. "Her childhood has passed right before our eyes," said Theresa Cedillo, "spent in hospitals and doctors' offices, not in parks and with little friends. The trauma of the sheer human suffering she endures every day is beyond explanation and understanding, filling us with overwhelming anguish."

For the next five days, Sylvia Chin-Caplan called upon a series of experts to connect the dots—to make sense of what had happened to Michelle Cedillo. First to testify was Vas Aposhian, an eighty-one-year-old professor of molecular and cellular biology at the University of Arizona and a professor of pharmacology at the University of Arizona School of Medicine. Aposhian said autistic children had less mercury in their hair than nonautistic children because they had a "mercury efflux disorder": they simply couldn't rid themselves of mercury. (Aposhian's claims were in direct contrast to Boyd Haley's claims that autistic children had more mercury in their hair than nonautistic children.) Aposhian said that when autistic children were treated with chelating agents, large amounts of mercury poured out of their bodies in the urine.

Next up was Arthur Krigsman, who displayed a poster he had recently presented at a national meeting. The poster told the stories of children whose symptoms were just like Michelle's. Krigsman said he was certain that Michelle's intestinal symptoms were caused by a long-standing, destructive measles infection caused by MMR. Chin-Caplan asked him how he could be so sure. Krigsman, pointing to his poster, said, "[It's] the same patients, the same bowel findings, and the same findings of measles virus genome. The pattern of inflammation that I was seeing was consistent with a viral infection."

Next was Vera Byers, an immunologist. Byers was impressed by Michelle's reaction to MMR. "[Michelle] had received her MMR vaccination at age fifteen months," she said. "Then a very colorful and dramatic set of circumstances occurred. Seven days later she developed a high fever, which essentially lasted for two weeks." Byers was confident that Michelle's reaction was caused by a "dysregulated immune system." Chin-Caplan asked, "Did you come up with potential possible causes?" "Yes," said Byers. "I find that both the thimerosal that she had received in her prior injections before MMR and the one or two [injections of thimerosal] subsequent to MMR were both responsible."

Through her experts, Chin-Caplan had tried to provide an explanation for Michelle Cedillo's autism. Vas Aposhian said Michelle was less able to rid mercury from her body than other children. Vera Byers said mercury from vaccines had damaged Michelle's immune system. And Arthur Krigsman explained that with her immune system damaged, Michelle was unable to prevent measles vaccine from damaging her intestinal lining. Only one piece was missing. How did immune suppression caused by mercury and long-standing intestinal inflammation caused by MMR add up to autism?

Marcel Kinsbourne, a professor of pediatrics at the Hospital for Sick Children in Toronto, provided the final piece of the puzzle. "The measles vaccine virus was able to access the brain, invade neurons without killing them [and] evoke a vigorous [immune] response," said Kinsbourne. "[As a consequence] the inflammation that resulted from that response damaged critical circuits in the brain." In short, measles vaccine virus had invaded Michelle's brain and caused autism. In support of his theory, Kinsbourne pointed to Unigenetics' detection of measles virus in the spinal fluids of autistic children. With that, Sylvia Chin-Caplan rested her case. Now it was Vincent Matanoski and the defense team's turn. They started by questioning the expertise of Aposhian, Byers, and Kinsbourne.

Vas Aposhian had testified that autistic children had greater quantities of mercury in their bodies than nonautistic children. Cross-examination of Aposhian revealed that the study on which he based his theory had been clearly refuted by other investigators. Further, Aposhian was not a toxicologist, geneticist, clinician, immunologist, or virologist; he had never taken care of a child with autism; he had never published a paper about thimerosal toxicity; and he had spent most of his time testifying in court on behalf of plaintiffs. But Aposhian's most embarrassing moment came after he described the Minamata Bay disaster. On cross-examination, Aposhian had to agree that despite this massive mercury poisoning, young children had not appeared to develop autism. Further, he was forced to admit that studies of mercury poisoning involved doses of mercury far greater than those contained in vaccines. Backed against the wall, Aposhian struck out when the examiner asked the next logical question: "Would you agree that any substance is either toxic or nontoxic based upon the dose?" "No," said Aposhian. "This is an ancient form of quotation that until recently we taught in medical schools, and in undergraduate school, and in graduate school. No longer can we use that ancient saying, and it's very ancient. This is the year 2000; it's not the year 1000 B.C." Aposhian concluded that "we no longer believe that the dose makes the poison."

Vera Byers, who said thimerosal had caused a "dysregulation" of Michelle's immune system, also had her credibility challenged. Byers had described herself as a member of the faculty of the University of Nottingham and later the University of California at San Francisco (UCSF) as well as a member of the clinical team "that got Embrel approved." (Embrel is a drug used to treat diseases like rheumatoid arthritis and psoriasis.) But Byers wasn't on the faculty at either Nottingham or UCSF, and her name never appeared on the Biologics License Application to the FDA for the approval of Embrel. Further, although Byers had claimed to be a toxicologist—indeed, her testimony centered on the supposition that thimerosal harmed the immune system—she had no formal training in toxicology, saying she had only taken courses in medical school. When questioned by Matanoski, she didn't know the chemical structure of thimerosal or the molecular weight of mercury, arguing she had relied on Vas Aposhian to know those things. (Aposhian wasn't a toxicologist either.) Finally, Byers hadn't taken care of patients in more than fifteen years, but instead had been spending all of her time running a consulting company called Immunology Incorporated, which provided testimony for personal-injury lawyers on toxins in the environment.

Arthur Krigsman, also an important witness for the petitioners, had claimed that the intestines of children with autism were studded with measles vaccine virus, as detected by both Andrew Wakefield in his laboratory and John O'Leary at Unigenetics. On cross-examination, Matanoski's colleagues didn't challenge what Krigsman had said. They just had him say it again, thereby setting him up for the surprising testimony of two British researchers a couple of days later.

Matanoski's final target was Marcel Kinsbourne, who had explained that measles vaccine virus caused autism by entering the brain and damaging nerve cells. But Marcel Kinsbourne wasn't a virologist. When questioned about how measles vaccine virus left the intestine and entered the brain, he struggled. Further, Kinsbourne hadn't cared for children in more than seventeen years. Chin-Caplan's choice of Marcel Kinsbourne as her autism expert would soon come back to haunt her.

Aposhian, Byers, and Kinsbourne had one thing in common: all had spent the past decade or more of their careers as professional witnesses for plaintiffs. Vincent Matanoski later commented on the petitioners' choice of experts when he addressed the judges: "Ask yourself on the credibility of witnesses where they're coming from. Is their place of business the hospital, or is it the courtroom? Do they get paid to testify, or do they testify to get paid?"

• • • •

MATANOSKI'S DEFENSE TEAM THEN PROVIDED SEVERAL EXPERTS who refuted the theory that MMR had caused Michelle Cedillo's autism. The first was Eric Fombonne.

Fombonne had trained at the Maudsley Hospital and the Institute of Psychiatry in London under Michael Rutter, one of the founders of child psychiatry as a scientific discipline. At the time of his testimony he was the head of the Division of Child and Adolescent Psychiatry and the director of the Autism Spectrum Program at Montreal Children's Hospital, as well as a professor of psychiatry at McGill University. Specializing in children with autism, Fombonne diagnosed more than 300 new cases every year. During his twenty-three-year career, Fombonne had published more than 160 research papers and 34 book chapters. He was on the editorial board of the *Journal of Psychology and Psychiatry*, an associate editor of the *Journal of Autism and Developmental Disorders*, and the president of the scientific committee for the International Meeting for Autism Research, the premier gathering for researchers in the field. Eric Fombonne was, in short, one of the world's leading experts on autism.

Fombonne contended that Michelle Cedillo's symptoms of autism appeared well before she had ever received MMR. To prove it, he showed the court a series of videotapes taken by her parents. One was from August 30, 1995, during Michelle's first birthday party. "Before we look," said Fombonne, "it's useful to try to portray in your mind what is typical of the first birthday party. So you will see there is cake, there is a gift, people sing. What you would expect from the child [is that] the child would be excited; there would be pleasure on the face of the child. If a child is called he would [pay attention] to the name. He would have a lot of interactions with people around. There would be a lot of showing, pointing, or gestures used to communicate. And if not words, you would at least hear babble." But that's not what Eric Fombonne saw when he watched the videotape of Michelle Cedillo. He focused on the segment of the tape where everyone was singing. "You need to pay attention to 'Happy Birthday, Michelle, Happy Birthday, Michelle,'" said Fombonne. "Twice or three times you will see that when she's spoken to, she doesn't [pay attention] at all. She's not [paying attention] to the face, she's not looking, she's not responding. There is no gesture, no pointing, no showing. We don't hear any babble. Her facial expressions are restricted and reduced. She doesn't join in when there is excitement." Fombonne was certain that after watching this tape, any autism expert would say that Michelle was autistic. The videotape of Michelle Cedillo's first birthday party was made four months before she received the MMR vaccine.

Fombonne later examined a videotape from December 17, 1995, when Michelle was fifteen months old, one month before she received MMR. "There is no word at all which is uttered by [Michelle] during this sequence," said Fombonne. "We just heard a few babbling sounds which are, again, guttural: not directed to others, directed to herself. We see flapping movements of the hand." Fombonne pointed to a scene where Michelle was handed several balls. "With the balls, which were a gift from her grandpa, she doesn't do anything. She doesn't explore them or play with them in any sort of way. Any child her age should play with toys she's given. Soon you are going to see something that is even more typical of autistic behaviors. Children with autism often have these very typical hand and finger movements whereby they move their fingers in their visual field like this." Fombonne feigned fascination as he slowly moved his hand, fingers spread widely, across his face. "They are absorbed by this," he said, pointing to the tape. "You are going to see this in six seconds. When her mother calls her, she doesn't respond. Rather, she engages in this stereotypical hand movement." Fombonne concluded his presentation of the videotapes: "This clearly suggests that [Michelle's] abnormal development occurred much before the MMR injection." Asked whether he was sure of his diagnosis, Fombonne didn't hesitate. "This set of findings based on video analysis is very consistent with autistic spectrum disorder," he said. "I have no doubt in my mind."

The testimony of Eric Fombonne refuted the petitioners' claim that Michelle's autism occurred after she had received an MMR vaccine. Chin-Caplan's only hope was to challenge the videotape observations point by point. But her autism expert, Marcel Kinsbourne, had never diagnosed a patient with autism, hadn't seen any patients in almost twenty years, and hadn't seen Michelle's behavioral abnormalities on the videotapes. As a consequence, when Sylvia Chin-Caplan cross-examined Eric Fombonne, she never once questioned his observations or conclusions; nor did she bring back Marcel Kinsbourne to challenge him.

· · · ·

THE NEXT WITNESS TO BLOW UP THE THEORY OF THE PETITIONERS was Stephen Bustin, a molecular biologist. Bustin was the chair of Molecular Science at the University of London and one of the first scientists to use PCR, the technique used by Unigenetics to detect measles virus in children with autism. In 2000, Bustin wrote the definitive paper, published in a premier biomedical journal, *Nature Protocols*, on how to perform PCR. This paper had been cited by other researchers more than 1,000 times, a remarkable number. More recently, he had written *A to Z of Quantitative PCR*, often referred to by molecular biologists as the bible of PCR. In the past five years alone, Bustin had written nine book chapters about PCR. For his expertise, he had been honored by election into the prestigious Royal Academy of Medicine. Stephen Bustin was arguably the world's expert on PCR.

When Andrew Wakefield declared MMR vaccine might cause autism, his claim was based in large part on finding measles virus genes in the intestines and spinal fluids of autistic children. Although Wakefield initially tested these samples in his own laboratory, he eventually sent them to Unigenetics in Dublin. So from January through May 2004, Stephen Bustin visited Unigenetics.

When Bustin examined Unigenetics' methods, he found something that didn't make sense. Unigenetics had taken samples from Andrew Wakefield and tested them using a type of PCR capable of detecting ribonucleic acid (RNA). Since measles virus contains only RNA, this was the only method that should have detected it. But on at least one occasion, Unigenetics detected measles genes using a type of PCR that could detect only deoxyribonucleic acid (DNA), and measles virus doesn't contain DNA. Clearly, something was wrong. "I have very little doubt that what they are detecting is a DNA contaminant and not measles virus," he said. Then Bustin found where the contamination was coming from. "One of the peculiar things that we noticed when we went to their laboratory," said Bustin, "was that next to their PCR instrument was a room which was labeled 'Plasmid Room.' [Plasmids are small circular pieces of DNA that are grown in bacteria.] Obviously if you have hundreds of millions or thousands of millions of bacteria, each containing tens of hundreds of copies of DNA, you've got a massive potential for DNA contamination. So you never want to have any plasmid DNA anywhere near your laboratory where you're doing the PCR. Once you've got DNA contamination, it persists for years and it gets into everything. If you're handling bacteria or you're handling plasmids, it gets into your hair, on your hands, [and] on your clothes."

Bustin found other problems with Unigenetics. Typically researchers run samples in duplicate or triplicate so that a specimen can be double- or triple-checked—in this case, to make sure it really does contain measles virus genes. Unigenetics ran its specimens in duplicate, occasionally finding measles genes in one duplicate sample but not the other. Despite this inconsistency, the company still sent out reports claiming the presence of measles genes. Also, researchers always check themselves with positive and negative controls: positive controls contain known amounts of measles virus RNA, and negative controls contain no measles virus RNA. But in the hands of Unigenetics, the positive controls were occasionally found to be negative and the negative controls were found to be positive. This suggested that something was critically wrong with the company's testing procedures. "I do not believe that there is any measles virus in any of the cases they have looked at," Bustin concluded.

Usually medical laboratories, which perform services critical to the care of patients, are accredited by a central licensing board. This wasn't the case with Unigenetics. "Was Unigenetics ever accredited?" Bustin was asked. "No, they were not," he replied. Then, "Could this be part of the reason some of these problems weren't detected earlier?" "Yes," said Bustin. "I'm sure that is the case. [Pathologists] tried to recruit Unigenetics into a quality control program, which involves various laboratories in Europe and the United States. Unigenetics refused to take part in this. So there was never any independent quality assessment made of any of the work that was carried out by Unigenetics." Bustin had known about problems with Unigenetics for more than three years, but because his findings related to litigation pending in the United Kingdom, he wasn't allowed to comment to the press or the public. Now, during the Omnibus Autism Proceeding, he could finally speak out about what he had found. "It has been incredibly frustrating," he told a reporter. "For three years we have been unable to reveal our findings. Now I want to get the message out about the O'Leary-Wakefield research. There's nothing in it."

Vas Aposhian, Vera Byers, Arthur Krigsman, and Marcel Kinsbourne had all said that their belief that MMR caused autism was based on Unigenetics' finding measles virus in the intestines and spinal fluids of autistic children. When the problems with Unigenetics were revealed, they each sat quietly, saying nothing.

Unigenetics Laboratories is no longer in business.

• • • •

WHEN ANDREW WAKEFIELD TESTIFIED IN FRONT OF DAN BUR-ton's congressional committee in 2002 that he had found measles virus in the intestines of more than 150 autistic children, Unigenetics had done all of the testing. Indeed, John O'Leary, the owner and operator of Unigenetics, had sat next to him, supporting his claim. But Wakefield's first paper—the one published in the *Lancet* that had started the controversy—wasn't based on tests performed by Unigenetics. Rather, Wakefield's *Lancet* paper was based on tests performed in his own laboratory. Andrew Wakefield didn't do these tests himself; a research assistant did them. His name was Nicholas Chadwick, and he was the next to testify.

Beginning in 1996, Chadwick was in the operating room during the collection of both intestinal biopsies and spinal fluids from autistic children. "My role was to take the material, bring it to the lab, and then look for evidence of measles RNA," he said. Although Chadwick's videotaped testimony lasted for only one hour, it was devastating to the petitioners.

Q: Did you personally test the gut biopsy samples for measles RNA?

A: Yes.

Q: What tests did you perform?

A: A PCR test, a polymerase chain reaction.

Q: What results did you [find] from the gut biopsy materials for measles RNA?

A: They were all negative.

Q: They were all negative?

A: Yes. There were a few cases of false positives. [But] essentially all the samples tested were negative.

Q: Did you personally test cerebrospinal fluid samples from autistic children in the lab?

A: Yes, I did. Again, they were all negative.

Q: Did you inform Dr. Wakefield of the negative results?

A: Yes. Yes.

Q: You also state in your affidavit that Dr. Wakefield was aware of all of your negative results when he submitted his paper, which was published in 1998 to the *Lancet*.

A: Yes, that's correct.

Q: Why wasn't your name on the paper I just referenced?

A: I asked for my name to be taken off anything that related to PCR data because I wasn't comfortable with the quality of the data.

Nicholas Chadwick testified that Andrew Wakefield had not only ignored data that disproved his contention, but he had also knowingly falsified them. If true, this revelation showed Wakefield had crossed the line from ill-conceived, poorly performed science to fraud.

Wakefield's willingness to misrepresent data wasn't new. Before he turned his attention to autism, Wakefield believed measles vaccine caused Crohn's disease. At a national meeting, he claimed that people with Crohn's disease had higher levels of measles antibodies in their bloodstream than people without the disease. One problem: David Brown was sitting in the audience. Brown, an internationally recognized measles expert, directed the WHO's measles laboratory in the United Kingdom. He knew people with Crohn's disease didn't have higher levels of measles antibodies because he was the one who had performed the tests. When Brown stood up to challenge Wakefield's statement, Wakefield had little choice but to withdraw it.

On cross-examination—even though their statements were devastating—neither Sylvia Chin-Caplan nor Thomas Powers challenged Stephen Bustin's claim that Unigenetics Laboratory was unreliable or Nicholas Chadwick's claim that Andrew Wakefield had falsified data. They simply asked a few perfunctory questions and got them off the witness stand as quickly as possible.

• • • •

THE THREE JUDGES IN CHARGE OF THE OMNIBUS AUTISM PROceeding aren't expected to reach a final verdict on whether vaccines might cause autism until 2009.

CHAPTER 9

Science and the Media

A good newspaper is never quite good enough, but a lousy newspaper is a joy forever.

—Garrison Keillor

When parents became concerned that vaccines had caused their children's autism, scientists responded by performing a series of epidemiological studies. All showed the same thing: vaccines weren't at fault. But despite the singular, consistent, reproducible, and clear results of these studies—and consequent reassurances from national and international health groups—many parents remain fearful. Why? Why has there been such a deep and persistent rift between the science that exonerated vaccines and the public's understanding of that science? Indeed, when people hear the word *vaccines*, one of the first things they think of is autism.

· · · ·

The public learns about science from lawyers, politicians, doctors, and scientists, most of whom filter their information through the media. Unfortunately, the motivations of scientists who perform studies differ from those in the media who describe them: one wants to inform, the other to entertain.

On August 7, 2005, Tim Russert of NBC's *Meet the Press*
examined the case against mercury in vaccines. Few journalists
were more respected than Russert, a serious and thoughtful
man whose programs were consistently praised for their excellence. If anyone could fairly review the subject of vaccines and
autism, it was Tim Russert. But Russert succumbed to the journalistic ethic of the time. He invited two people onto his program: Harvey Fineberg, president of the Institute of Medicine
and former dean of Harvard's School of Public Health, and David Kirby, the journalist who wrote *Evidence of Harm*. By struc-
turing his program as a contest, Russert eliminated any chance
to inform his viewers. Here's what went wrong.

At the time of the Russert interview, four of the eight studies
that had exonerated thimerosal had already been published, and
although it was still a source of controversy in the media and in
court, the vaccines-cause-autism hypothesis was no longer via-
ble among scientists. If Russert had genuinely wanted his viewers to understand the issue, he would have interviewed Harvey
Fineberg only. Fineberg would have first explained what epidemiological studies are, how they work, and why they are the
best way to determine whether one thing causes another. Then,
he would have described how scientists would select two groups
of children that are alike in all respects except one—their receipt of thimerosal. He would have described how the size of a
study determines its statistical power: small studies may be sensitive enough to implicate thimerosal as a cause of autism in
only one in 100 children, but larger ones could detect it in one in
1 million children. And he would have explained how scientists
would randomly select children who had or hadn't received
thimerosal so as not to bias the results. By interviewing Harvey
Fineberg only, Tim Russert would have given his viewers a better understanding of the strength and consistency of the epidemiological studies that had been done. The obvious problem
with offering such a tutorial on network television is that it's
painfully boring. So Russert did what almost every other journalist who writes or talks about science does: he set it up as a

controversy with no intention to resolve it. He pitted an entertaining journalist with a background in public relations against a careful, thoughtful scientist. One man was made for television; the other wasn't. If questioned, Russert would have probably offered the journalistic mantra of "balance": in order to flesh out a controversy, he had to provide both sides of the argument. But there's a difference between balance and perspective. A more accurate balance—and a fairer representation of the prevalent view—would have been to have interviewed 1,000 scientists who, having reviewed the evidence that exonerated vaccines, represented one side, and a single scientist, like Mark Geier, who wasn't convinced. But it also wouldn't have been a very good television program. So Russert opted for a dramatic, one-on-one twenty-minute confrontation.

Harvey Fineberg remembers his appearance on *Meet the Press*. "It's a little like the [Samuel Johnson] metaphor about hanging," he recalled. "It concentrates the mind." The result was predictable. Fineberg did a wonderful job describing the science that had exonerated mercury, a nearly impossible task given the few moments he had to do it. Kirby, on the other hand, dismissed Fineberg's epidemiology with a wave of his hand and alluded to exciting new findings by researchers studying laboratory cells and experimental mice (such as those performed by Richard Deth and Mady Hornig). Fineberg didn't have the time and Russert didn't have the interest in hearing about how studies in the laboratory could never be as valuable as studies in hundreds of thousands of children. But that didn't matter. It was great theater. Tim Russert had taken a boring subject like epidemiology and transformed it into an exciting confrontation: a war between a young journalist fighting for the rights of a disenfranchised group and a mainstream scientist who offered only epidemiological studies and their statistical results.

The Russert interview on *Meet the Press* shows why it is so difficult to educate the public about science. For his network, Russert's success is judged in large part by the size of his audience and his ability to sell advertising; a show that carefully de-

Harvey Fineberg (left) *and David Kirby square off during a* Meet the Press *interview with Tim Russert, August 7, 2005 (courtesy of Getty Images).*

fines how epidemiological studies are performed won't accomplish either. On the other hand, a confrontation between an enthusiastic muckraking journalist and a scientist who represents a faceless giant like the Institute of Medicine will. Judea Pearl, a professor of computer science at UCLA, said it best: "Journalists cannot simply pour gasoline into the street and pretend they bear no responsibility for the inevitable explosion." The need to sell advertising—to be vivid, dramatic, and interesting—stands in constant opposition to the public's understanding of science.

Like most journalists without a scientific background, Tim Russert had little knowledge of the workings of science, so he didn't focus on science. He focused on people—a scientist and a journalist. People are much more interesting than science. In the case of vaccines and autism, it isn't hard to find scientists on both sides of the debate. But, in truth, it isn't hard to find scientists on both sides of any issue, independent of whether it's a debate. For example, to take this notion to an extreme, the science program *Nova* occasionally airs shows describing plate tectonics, the mechanism by which the earth's surface moves.

The concept of plate tectonics assumes the earth is round. But if the producers of *Nova* ever wanted to make plate tectonics more interesting, they could include a scientist who disagrees, claiming the earth is flat. Indeed, several scientists belong to the Flat Earth Society, an active, engaging group dedicated to "deprogramming" the masses since the sixteenth century. The mission statement of the Flat Earth Society would make for great television: "For centuries, mankind knew all there was to know about the shape of the Earth. It was a flat planet, shaped roughly like a circle. Then, in the year of our Lord fourteen-hundred and ninety-two, it all changed. Christopher Columbus, using an elaborate setup involving hundreds of mirrors and a few burlap sacks, was able to create an illusion so convincing that it was actually believed he had sailed around the entire planet and landed in the West Indies. As we now know, he did not." *Nova* has chosen not to contradict the science of plate tectonics with the Flat Earth Society's contention that the earth is flat. That's because the earth is round. But the notion that vaccines cause autism has also been clearly disproved. Still, the issue is reported as a controversy.

• • • •

ANOTHER INHERENT BIAS OF JOURNALISTS IS THAT THEY SEE themselves as defenders of the weak against the powerful. In the late 1800s, Finley Peter Dunne, an editorial writer for the *Chicago Post*, stated that journalists should "comfort the afflicted and afflict the comfortable." In the vaccine-autism story, the media cast children with autism as the *afflicted* and pharmaceutical companies, public health officials, doctors, and scientists as the *comfortable*. When Tim Russert picked David Kirby for his show, he didn't pick him because he was an expert on autism (Kirby had never diagnosed or treated a patient with autism); or because he was an expert on mercury poisoning (Kirby wasn't a toxicologist); or because he was an expert on vaccines (Kirby wasn't an immunologist, virologist, or microbiologist); or because he was an expert on epidemiological studies (Kirby had never performed or published an epidemiological study). David Kirby had no specific expertise in any aspect of the thimerosal-autism debate. But Kirby had written a book claiming that public health officials, knowing that thimerosal had harmed the unsuspecting children of America, had done everything they could to cover it up. David Kirby was a journalist's dream—bright, articulate, young, and attractive; he was the plucky little guy willing to take on the evil big guy. Harvey Fineberg—cast unfairly in the role of Goliath—had been put in an impossible position before he had ever stepped onto the *Meet the Press* set.

Arthur Allen, the journalist who wrote *Vaccine: The Controversial Story of Medicine's Greatest Lifesaver* and later debated David Kirby in San Diego, laments the manner in which the public is educated about science. "Every time some schmo who people have seen on television buys into [the vaccine theory] for whatever reason, it just keeps getting more of a life," he said. "I find it distressing to see people like [David Kirby] being given more authority by the media than the CDC. There are things definitely worth investigating, and bad things get done by people. But that's not always the narrative. That's not what journalism is supposed to do. I still like the 'comfort the afflicted and afflict the comfortable' line about journalism, but that doesn't mean misrepresenting things."

Tim Russert wasn't alone in his choice to tell the vaccine-autism story as a David versus Goliath tale. Virtually every major newspaper, radio station, and television network told the story the same way. (The David-versus-Goliath theme isn't confined to television and radio; movies, too, often portray issues in medicine and science as a confrontation between people willing to take on the rich and powerful who are bent on destroying them. Movies like *Lorenzo's Oil*, *Erin Brockovich*, *Silkwood*, and *A Civil Action* all described how the little guy can take on the big guy and win. Americans love these stories. All of these movies were riveting, and all did well at the box office.) Although the David-versus-Goliath theme is compelling, journalists typically miscast the players in the vaccine-autism controversy. When doctors and scientists stand in front of the media to dismiss the contention that vaccines cause autism, they *are* representing the little guy. In this case, the little guy is the autistic child subjected to harmful therapies or denied potentially life-saving vaccines. .

. . . .

NOT ALL NEWSPAPER REPORTERS CUT THEIR MORALS TO FIT THE style of the time. On June 25, 2005, Gardiner Harris and Ana-had O'Connor wrote a lengthy article that appeared on the front page of the *New York Times*. The article wasn't equivocal. Harris and O'Connor presented the epidemiological evidence disproving the notion that thimerosal caused autism; gave short shrift to the fringe scientists who disagreed; and criticized parents who subjected their children to radical, unproven, and potentially dangerous therapies. They offered a perspective based on good science, and they were hammered for it. Daniel Schulman, an assistant editor for the *Columbia Journalism Review*, denounced the article as shamefully one-sided. "The story cast the thimerosal connection as a fringe theory without scientific merit, held aloft by angry, desperate parents," said Schulman. "The notion that supporters of the theory were disregarding irrefutable scientific findings was an underlying theme, drilled home several times. Readers were left with little option but to believe that the case against thimerosal was scientifically unsound. Several reporters I spoke with who have covered the thimerosal controversy described the *Times* story as a 'hit piece.'" Schulman didn't stop there. He praised journalists such as Myron Levin of the *Los Angeles Times*, Dan Olmstead of United Press International, and Craig Westover of the *St. Paul Pioneer Press* for their bravery in taking on powerful establishments. Westover specifically was praised for standing up to vitriolic comments on his blog following the mercury chelation death of Tariq Nadama. One blogger had written: "They finally did it, Mr. Westover. They killed a little boy trying to get that satanic mercury out of his little body. You have some blood on your hands. Like it or not, you do. There has been no autism epidemic and thimerosal doesn't cause autism. I hope the parents of this boy point the finger at you and scream murder."

Craig Westover considered his response. "I really do try to walk a middle line on this," he said. "You have to go out and investigate this and be able to come to some sort of conclusion. Not definitely that thimerosal does or does not cause autism, but you have to come to the question of whether this theory is plausible or not. Otherwise, I think you're doing a disservice to your reader." Westover concluded that the thimerosal-autism theory was plausible. Later, he responded to the blogger: "This is the risk of a sin of commission," he wrote, "and one I considered long and hard before I wrote my first article on this topic. I will stand on what I believe and accept the risk and the consequences if I am wrong."

When Craig Westover wrote his articles, he knew that he was protected by the legal treatise of "absence of malice," which states that a journalist cannot be held accountable for false statements unless it can be proved that he made them knowingly and maliciously. It's a high bar. So it is unclear what Westover meant when he said he would accept the consequences if he were found to be wrong. He was wrong and Tariq Nadama is dead. Did Westover mean he would call up and apologize to the parents of Tariq Nadama? Or stand beside Roy Kerry, the physician responsible for Tariq's death, after he had been indicted for involuntary manslaughter? Or issue a retraction in his newspaper? Certainly since he made the statement—during which time study after study continued to refute a notion he believed to be true—Craig Westover has done nothing to show he has accepted the consequences of his mistake. It's easy to claim responsibility when you know you'll never be held accountable.

. . . .

ANOTHER OBSTACLE TO THE PUBLIC'S UNDERSTANDING OF SCIence is that journalists love scientific mavericks. "Journalists typically cover the news," says Steven Pinker, a professor of psychology at Harvard University and the author of *The Stuff of*

Thought, "with the finding that upsets the apple cart rather than the consensus."

In the past, vaccine stories have had a remarkable sameness; doctors talked about how vaccines saved lives, and scientists talked about the wonder of creating them. Andrew Wakefield and Mark Geier were a breath of fresh air, taking a boring story and making it controversial, full of scandal and intrigue. They stood apart from conventional thinking; apart from government agencies, advisory committees, and pharmaceutical companies; and apart from the physicians' mantra that vaccines were safe. They were among a precious few who appeared willing to speak truth to power. Long after Andrew Wakefield's notion that MMR caused autism had been disproved, Melanie Phillips of London's *Daily Mail* continued to represent him as a hero. "There are very powerful people who have staked their entire reputations and careers on proving Andrew Wakefield wrong," wrote Phillips, "and they are willing to do almost anything to protect themselves. While Mr. Wakefield is being subjected to a witch hunt, and while the parents of affected children are scandalously denied legal aid to pursue the court case which may have finally brought to light the truth about MMR, those powerful people in the medical establishment are continuing to misrepresent the evidence." Phillips was arguing that the history of medicine was studded with scientists who stood outside the system, were ridiculed for it, and were eventually proved right. Wakefield, according to Phillips, was no different.

In fact, all scientists, if they are to be successful, are iconoclastic. "Science," says George Johnson in his review of Freeman Dyson's *The Scientist as Rebel*, "is an inherently subversive act. Whether overturning a long-standing idea or marshalling the same disdain for received political wisdom, the scientific ethic— stubbornly following your nose where it leads you—is a threat to establishments of all kind." Said J. B. S. Haldane, a British geneticist and evolutionary biologist: "Beware of him in whom reason has become the greatest and most terrible of passions."

Scientists, bound only by reason, are society's true anarchists. Indeed, some of the greatest advances in medicine have been made by scientists who initially stood alone. For example, Barry Marshall, working at the Royal Perth Hospital in Western Australia, argued that an unusual bacterium called *Helicobacter pylori* caused stomach ulcers. No one at the time believed that bacteria could survive the harsh acid produced by the stomach, much less reproduce and cause disease. But Marshall was so convinced by his findings that he swallowed a Petri dish full of the bacteria, later developing severe inflammation in his stomach. In 2005, Barry Marshall won the Nobel Prize in Medicine. Following up on the work of radiation biologist Tikvah Alper and mathematician John Stanley Griffith, microbiologist Stanley Prusiner, from the University of California at San Francisco, argued that proteins alone could cause infections. Scientists knew that bacteria, parasites, viruses, and fungi were infectious because each of these organisms contained genetic material that allowed them to reproduce. Prusiner's notion that a single protein could cause an infection was heretical. But Stanley Prusiner was right. And his proteinaceous infectious particles (prions) were later found to be the cause of mad cow disease. Prusiner won the Nobel Prize in Medicine in 1997. And Albert Einstein, before he described his theory of relativity, proposed that light was composed of tiny particles. No one believed him. Later, researchers found that light rays were redirected by the gravitational pull of the sun, proving Einstein right. Einstein won the Nobel Prize in Physics in 1921.

Marshall, Prusiner, and Einstein had several things in common: they were decades ahead of their time, their findings were initially ignored, they stood their ground, and other scientists eventually proved them to be right. And, like Andrew Wakefield, all were considered to be rogue scientists and were criticized for their hypotheses. But unlike Andrew Wakefield and his proposal that MMR caused autism or Mark Geier and his proposal that thimerosal caused autism, all of these Nobel Prize–winning scientists had their work confirmed by other investigators—redundantly. For Wakefield and Geier, this hasn't been the case; despite years of study, many groups of investigators working on several different continents have failed to support their theories. In short, not all rogue scientists are good scientists. "History is replete with tales of the lone scientist working in spite of his peers and flying in the face of doctrines," wrote Michael Shermer, author of *Why People Believe Weird Things*. "Most of them turned out to be wrong and we do not remember their names. For every Galileo shown the instruments of torture for advocating scientific truth, there are a thousand or ten thousand unknowns whose 'truths' never pass muster."

. . . .

ANOTHER TRAP FOR JOURNALISTS IS THE LURE OF THE SINGLE study. After Andrew Wakefield published his paper in the *Lancet* claiming that MMR caused autism, journalists jumped at the chance to report his dramatic new finding. But scientific theories aren't proven by the number of journalists who write about them. Novel, and in this case shocking, claims are best proved by further study. That's because scientists, even excellent scientists working at prestigious institutions, often get it wrong—and it's not hard to publish bad science. "Obviously, we are all interested in the truth," said Arnold Relman, former editor of the *New England Journal of Medicine*. "But it's mostly what happens after a study is published that determines truth." For example, in 1981, Brian MacMahon and his colleagues suggested that coffee drinking could lead to pancreatic cancer. The study, which included extensive interviews with 400 cancer victims, was carefully performed, evaluated, and described; it was published in the *New England Journal of Medicine*, one of the best medical journals in the world. And MacMahon and his coworkers were from Harvard's School of Public Health, a premier institution. The media carried this story as fact, and for a little while people were more circumspect about their coffee drinking. But Brian MacMahon was wrong. Study after study failed to reproduce his results, and the notion that coffee caused pancreatic cancer faded away.

Sometimes the public is confused and disillusioned when a much-heralded study fails to survive closer scrutiny, believing that science cannot be trusted to get it right. "It is important to recognize the fallibility of science and the scientific method," wrote Michael Shermer. "But within this fallibility lies its greatest strength: self-correction. Whether a mistake is made honestly or dishonestly, whether a fraud is unknowingly or knowingly perpetrated, in time it will be flushed out of the system by lack of external verification." Unfortunately, some people are uncomfortable with the fluidity of science, looking for something immutable and certain. "When better information is available, science textbooks are rewritten with hardly a backward glance," says Robert Park, a professor of physics at the University of Maryland and the author of *Voodoo Science: The Road from Foolishness to Fraud*. "Many people are uneasy standing on such loose soil; they seek a certainty that science cannot offer. For those people the unchanging dictates of ancient religious beliefs, or the absolute assurance of zealots, have a more powerful appeal."

To best serve their readers, journalists should be skeptical of any scientific study that appears to break new ground. Following astronomer Carl Sagan's warning that "extraordinary claims require extraordinary proof," they should wait until the next round of studies confirms the initial one before unnecessarily frightening the public. This hope, of course, is fanciful. The lure of dramatic headlines, advertising dollars, and ratings is far stronger than the desire to avoid scaring the public with an unconfirmed study.

• • • •

PUBLIC OPINION IS ALSO INFLUENCED BY PARENT ADVOCACY groups and the public relations firms that work for them. During the thimerosal debate, Fenton Communications, hired by J. B. Handley's Generation Rescue, constantly lobbied the media with press releases touting laboratory studies (like those of Boyd Haley, Richard Deth, and Mady Hornig) that appeared to contradict epidemiological studies. Fenton used the same strategy during the breast implant controversy—a strategy pioneered by a legendary public relations firm working for the tobacco industry in the 1950s.

In 1953, tobacco companies hired the most powerful public relations firm in the United States, Hill and Knowlton. The firm's job was to seed doubt about the validity of epidemiological studies that clearly showed cigarette smoking caused lung cancer—to make the public believe the case against tobacco was a medical controversy. In the same way Robert F. Kennedy Jr., prominent politicians, and celebrities stepped forward to support the notion that thimerosal caused autism, Hill and Knowlton engaged entertainer Arthur Godfrey to defend cigarettes. During his weekly television variety show, Godfrey said, "I smoke two or three packs of these things every day. I feel pretty good. I don't know; I never did believe that they did any harm."

Hill and Knowlton also used personal testimonials to trump epidemiological studies. Advertisements by R. J. Reynolds, the maker of Camel cigarettes, featured people who had taken their own thirty-day test to determine if cigarettes were harmful. Elana O'Brien, a real estate broker, said: "I don't need my doctor's report to know Camels are mild." (Dan Burton used the same strategy during his congressional hearings in which parent after parent testified that MMR caused autism, science be damned.)

Then Hill and Knowlton convinced Edward R. Murrow, the television journalist who hosted the program *See It Now*, to do a show on tobacco. (Murrow was a chain smoker, consuming sixty to seventy cigarettes a day.) As described by Allan Brandt in *The Cigarette Century*, "Hill and Knowlton got precisely what they had hoped for, an ambiguous conclusion noting that more scientific research would be needed to settle the question." Following Murrow's show, people were confused about the science proving cigarette smoking caused lung cancer, believing it to be a debate among scientists. "Hill and Knowlton had successfully produced uncertainty in the face of powerful scientific consensus," wrote Brandt. "So long as this uncertainty could be maintained, so long as the industry could claim 'not proven,' it would be positioned to fight any attempts to assert the regulatory authority of public health." (Both Arthur Godfrey and Edward R. Murrow later died from lung cancer.)

Fenton Communications can claim equal success. Most people would probably agree with the statement, "Vaccines might cause autism," despite the publication of sixteen epidemiological studies that show they don't.

. . . .

ANOTHER OBSTACLE TO THE PUBLIC'S UNDERSTANDING OF SCIENCE is that it is often explained by lawyers with much to gain from championing a particular case. During the MMR debate, Richard Barr launched a Web site and distributed written materials explaining exactly how MMR caused autism; he was constantly quoted in the media. The same has been true in the thimerosal debate. Kevin Conway, Thomas Powers, Clifford Shoemaker, Sylvia Chin-Caplan, and lawyers for plaintiffs in vaccine cases have either been quoted in the press or on their Web sites explaining how mercury causes autism. All of these lawyers have been wonderful advocates for their clients. They have done much to convince the press and the public (which contains potential jurors) of the rightness of their cause. But lawyers aren't scientists; they're not seeking some scientific truth. They're trying to win cases for those who are paying them to do it.

During the Omnibus Autism Proceeding, the job of the lawyers representing Michelle Cedillo had been particularly challenging. Because the science wasn't on their side, plaintiffs' lawyers resorted to several different strategies to win favor from the presiding judges. When Eric Fombonne reviewed videotapes of Michelle Cedillo, showing that her autism had begun long before receiving an MMR vaccine, he dealt a devastating blow to the petitioners. A reasonable response by Sylvia Chin-Caplan and the plaintiff team would have been to call their own experts to refute Fombonne's testimony, but the evidence against them was too strong. So the lawyers appealed to emotion. To refute Fombonne's testimony, Chin-Caplan called as her final witness Theresa Cedillo, Michelle's mother. Theresa talked about how hard it had been to deal with Michelle's illness and how, no matter what Eric Fombonne had said, she knew that her daughter was acting normally before she had gotten MMR. Chin-Caplan hoped the judges' understandable desire to help a child who was suffering would cause them to ignore the videotape evidence that had been so devastating.

The choice to have Theresa Cedillo as the final witness wasn't the only appeal to emotion by the petitioners. At the front of the courtroom was a podium for the lawyers to address the judges. When defense attorney Vincent Matanoski made his opening statement, he didn't move it. But every time the petitioners' lawyers addressed the court, they turned the podium around and addressed the audience, which consisted of journalists and parents. They knew that the Omnibus Autism Proceeding wasn't going to be the end of this. They hoped their case against vaccines would eventually spill over into state courts in front of juries, not federally appointed judges. In state courts, plaintiffs wouldn't be suing the federal government (which had a war chest of $2 billion); they'd be suing pharmaceutical companies (which had been on the hook for ten times that amount in recent medical-product litigation). And plaintiffs' lawyers knew that potential jurors would be influenced by how the media covered this particular trial. The awkward positioning and repositioning of the podium was a comical reminder of the conflicting interests of the defense and plaintiff teams: the former was trying the case at hand, the latter was trying to influence the media and the public for subsequent trials.

After Eric Fombonne's testimony, the only other strategy left to the petitioners' lawyers was to question how scientists know things. They never disputed the fact that at the time ten separate epidemiological studies had exonerated MMR or that five had exonerated thimerosal; rather, they disputed the reach of those studies. Scientists are only human, they reasoned, they can't know everything. Those studies certainly weren't large enough to prove that vaccines couldn't cause autism in one in several million children; no epidemiological study was that powerful. And 4 million children were born in the United States every year. Perhaps vaccines caused autism in only a handful—one of them could be Michelle Cedillo.

The plaintiffs' argument that there are things we cannot know harkens back to the celestial teapot analogy first described in 1952 by philosopher Bertrand Russell. "If I were to suggest to you that between the Earth and Mars there is a china teapot revolving around the sun in an elliptical orbit," said Russell, "nobody would be able to disprove my assertion provided that I were careful to add that the teapot is too small to be revealed even by our most powerful telescopes. But if I were to go on to say that *because* my assertion cannot be disproved, it is an intolerable presumption on the part of human reason to doubt it, I should rightly be thought to be talking nonsense." Certainly it is true that scientists can't know everything, that the scientific method has limits, and that epidemiological studies cannot detect extremely rare events. But to use these truths as a basis to claim that MMR and thimerosal caused autism, to build an industry based on mercury-binding therapies or chemical castration, and to sue the federal government and pharmaceutical companies for the harm they have caused is an unjustified and dangerous leap.

The celestial teapot argument hasn't played well in state courts. Although almost all lawsuits against vaccine makers must first go through the federal vaccine court, there are a few exceptions. One pertains to children injured before 1986, when the National Childhood Vaccine Injury Act was created. In these cases, plaintiffs can take their chances in civil court. Such was the case of Jamarr Blackwell, an autistic boy whose parents sued Sigma Aldrich. The judge, Stuart Berger, didn't buy the notion that vaccines caused autism in a small group of genetically susceptible children, recognizing the obvious flaw in the logic. "Indeed, if plaintiffs' theory was based on generally accepted scientific principles," wrote Berger, "the autism allegedly caused in this subgroup would not be a 'rare event.' Dr. Geier testified that 80 to 90 percent of the cases of autism occurring in the late 1990s were due to exposure to mercury in childhood vaccines. If that were true, and those cases presented themselves in the genetically susceptible subgroup, then epidemiological principles would dictate that a large proportion of the population would have that genetic susceptibility. Moreover, such an effect would have been detectable in epidemiological studies of the general population."

· · · ·

POLITICIANS, BY WEIGHING IN ON SCIENTIFIC DEBATES, HAVE ALSO confused the public. During the vaccine-autism controversy, Joe Lieberman, then a Democratic senator from Connecticut; John Kerry, a Democratic senator from Massachusetts; Christopher Dodd, another Democratic senator from Connecticut; and Robert F. Kennedy Jr., a member of the most famous Democratic political family in America, all warned of the danger of vaccines—warnings that appeared on Don Imus's national radio program and in full-page advertisements in the *New York Times* and *USA Today*. Why? Given the wealth of epidemiological studies clearly showing vaccines didn't cause autism, why did these politicians stand up and tell the press and the public they did? A cynical view would be that they were paid to do it. Many Democratic politicians receive healthy contributions from the Association of Trial Lawyers of America (now the American Association for Justice), one of the most powerful lobbies in Congress. But the vaccines-cause-autism chant wasn't sounded only by Democrats. Dan Burton, who held a series of hearings implicating vaccines as a cause of autism, is a Republican. So is Arnold Schwarzenegger, who was among the first governors to ban thimerosal-containing vaccines from his state. And perhaps the most persistent and effective fighter on behalf of the notion that vaccines have been harmful is Dave Weldon, a Republican congressman from Florida. Weldon introduced federal legislation to ban thimerosal from all vaccines, pressured CDC director Julie Gerberding to let parent advocacy groups direct autism research, and constantly questioned the motives and competency of the CDC's Immunization Safety Office. The vaccine-autism debate has been stoked by politicians on both sides of the aisle.

The more likely explanation for politicians' involvement in the autism debate is that they have been responding to their constituents—us. Or at least those of us who've been the loudest. Activist groups that are the best organized, best funded, and best connected are the ones most likely to gain political attention, and standing up against mercury isn't a very heavy political load to lift. Unfortunately, by constantly beating the drum that vaccines cause autism, these politicians have failed to serve those for whom they are responsible: the children in their districts and states. Their scaremongering has only encouraged some parents to subject their autistic children to potentially harmful therapies or to withhold vaccines that might save their lives. In the name of protecting children, these politicians have worked against them. It's been a disappointing parade to the congressional podium.

• • • •

SCIENTISTS ALSO FAIL IN EDUCATING THE PUBLIC ABOUT SCIENCE. That's because, for the most part, they're reluctant to do it. "The reluctance of scientists to publicly confront voodoo science is vexing," wrote physicist Robert Park. "While forever bemoaning general scientific illiteracy, scientists suddenly turn shy when given an opportunity to help educate the public by exposing some preposterous claim. If they comment at all, their words are often so burdened with qualifiers that it appears that nothing can ever be known for sure. This timidity stems in part from an understandable fear of being seen as intolerant of new ideas. It also comes from a feeling that public airing of scientific disputes somehow reflects badly on science. The result is that the public is denied a look at the process by which new scientific ideas gain acceptance."

Because most scientists are reluctant to educate journalists or to stand in front of television cameras, the education is left to scientists with other motives. These scientists fall into two groups: those who ignore data and those who overinterpret them. The first group is represented by Andrew Wakefield and Mark Geier. Because they were convinced that vaccines caused autism, Wakefield and Geier shoehorned their data and disregarded studies that contradicted their hypotheses. Both have paid a heavy price for this, having been marginalized in the scientific and medical community. Wakefield, under investigation in England, sought refuge in the United States; and Mark Geier's expert testimony has been thrown out of several different courtrooms. Both of these men had faith in their beliefs, even in the face of overwhelming data to the contrary. But Wakefield and Geier failed to recognize that science isn't about faith; it's about data. Eric Fombonne offered a rare glimpse into the mind of a good scientist in an exchange with Sylvia Chin-Caplan during the Omnibus Autism Proceeding. "Do you believe that autistic children do not have bowel problems?" asked Chin-Caplan. "Actually, I have no beliefs in general," responded Fombonne. "What I look at is the evidence."

But Wakefield and Geier are unusual. More common is the second group of scientists who seek out the media. This group is represented by Richard Deth and Mady Hornig, both of whom are excellent, well-respected, well-published researchers working at prestigious universities. Unfortunately, both committed the intellectual sin of overinterpreting their data. Deth had found that thimerosal altered an important metabolic pathway of cancer cells grown in laboratory flasks, and Hornig had found that thimerosal altered the behavior of a highly inbred strain of mice. Their studies were a far cry from proving that thimerosal caused autism in children. To be reasonable, Deth and Hornig could have presented their findings with the appropriate caveats (noting, for example, that cancer cells in flasks aren't brain cells in people and that mice aren't children). But they didn't. They stood in front of congressional committees and television cameras and declared that thimerosal caused autism. Robert Park has commented on the phenomenon of scientists who descend into foolishness. "Even eminent scientists," said Park, "have had their careers tarnished by misinterpreting unremarkable events in a way that is so compelling that they are thereafter unable to free themselves of the conviction that they have made a great discovery. If scientists can fool themselves, how much easier is it to craft arguments deliberately intended to befuddle jurists or lawmakers with little or no scientific background?"

. . . .

SCIENTIFIC INFORMATION IS SHAPED BY THE SCIENTISTS, LAWyers, and politicians who influence the media, as well as by the media themselves. But there is another influence, one that is arguably even more powerful: the culture in which scientific information is offered. The vaccine-autism controversy offers many examples of how our current culture distorts the perception of science.

CHAPTER 10

Science and Society

There is nothing to fear except
the persistent refusal to find out the truth.

—Dorothy Thompson

Science is influenced by society. In the fourth century B.C. two Greek philosophers, Plato and Aristotle, believed the earth was the center of the universe (geocentrism). By the Middle Ages, everyone believed it. Further support for the theory of geocentricism came from Christian biblical references, such as "The world is firmly established; it cannot be moved" (Psalm 93:1, Psalm 96:10, and Chronicles 16:30) and "[The Lord] set the earth on its foundations" (Psalm 104:5). As a consequence, geocentrism assumed the power of religious dogma.

In 1543, however, Copernicus, a Polish mathematician, challenged the notion of geocentrism by claiming the earth revolved around the sun, not the other way around. Few believed him. But in December 1610, one year after the invention of the telescope, the Italian astronomer Galileo proved Copernicus was right. Galileo showed that Venus exhibited a full set of phases similar to the moon, a phenomenon that could have been possible only if Venus, like the earth, rotated around the sun. Galileo published his findings in his *Dialogue Concerning the Two*

Chief World Systems. But Galileo's work didn't sit well with the Roman Catholic Church, and in 1633, the papacy accused him of heresy. The trial didn't last very long. Church officials ruled, "The proposition that the sun is in the center of the world and immovable from its place is absurd, philosophically false, and formally heretical, because it is expressly contrary to Holy Scriptures." The Church banned Galileo's offending book, forbade publication of his future works, and ordered him imprisoned for the rest of his life. But Galileo knew he was right; as he was led away from his Roman inquisitors, he muttered, referring to the earth: "Eppur si muove" (And yet it moves).

Galileo's science didn't fit the culture of his time, so he was denounced for it. Today is no different. On October 5, 1999, Dan Burton appeared on the CNN program *Talk Back Live* with Bobbie Battista. Burton believed the MMR vaccine had caused his grandson's autism, and he planned to use his position as the chairman of the Committee on Government Reform to prove it. During the broadcast, a doctor on the program challenged Burton by describing a recent study by British epidemiologist Brent Taylor that contradicted Burton's theory. Burton was incensed. "At the New Jersey Conference, Stop Autism Now, there were twelve hundred parents," he said. "And they were asked the question 'Do you believe that the autism of your child was caused by vaccines?' Seven hundred and fifty raised their hands!"

Burton considered common belief to be common wisdom. If most people believed vaccines caused autism, then vaccines caused autism. Later, when Brent Taylor presented his data in front of Burton's congressional committee, Burton denounced him in much the same way the Church had denounced Galileo. But instead of using Holy Scriptures to make his case, Burton used the weapon of his time: conflict of interest. He accused Taylor and others of being unduly influenced by the federal government and pharmaceutical companies. "We have been checking into all the financial records," said Burton, "and we are finding some possible financial conflicts." Burton, appealing to the prevalent notion that everyone is in someone's pocket, implied that because of these unseen influences, Taylor's study should be discarded. "We are slipping into a new form of darkness," wrote Steven Milloy, "one where it's popular, profitable and politically expedient to suppress science."

. . . .

IN A CULTURE DOMINATED BY CYNICISM AND HUNGRY FOR SCANdal, many people believe that doctors, scientists, and public health officials cater to a pharmaceutical industry willing to do anything—including promote dangerous vaccines—for profit. So it's not hard to appeal to the notion that pharmaceutical companies are evil. During the breast implant controversy, one patient advocate said, "First, let's get over the myth that just because Harvard or the Mayo Clinic or Yale says that something is correct, that it is correct. We know where their bread is buttered. We know who gives the funding. Manufacturers fund; scientists do the studies." Comments made during the vaccine-autism controversy were no different. "I'm a patriot," said Boyd Haley. "But the thing I find very discouraging about our government is that we're more interested in protecting the income of professionals and the pharmaceutical industry than in protecting the American people."

Current movies also reflect this sentiment. In *The Constant Gardener*, released in 2005, a pharmaceutical company makes an antibiotic that is highly effective against multidrug-resistant tuberculosis. When the drug is found to have a fatal side effect, the company buries its victims in a mass grave outside of town and kills others who know about the problem, including the sympathetic wife of a government official. In *The Fugitive*, released in 1993, a pharmaceutical company hires a one-armed man to kill a doctor (Richard Kimble) when he finds that one of the company's drugs, nearing FDA approval, causes fatal liver damage. Neither the screenwriters nor the public considered these two scenarios implausible. Viewers were perfectly willing to believe that pharmaceutical companies hire hit men to kill people.

To some extent, pharmaceutical companies have brought this upon themselves. Twenty years ago, direct-to-consumer advertising of prescription medicines was uncommon. Now television viewers encounter a barrage of advertisements from pharmaceutical companies showing that medicines can be miraculous; people with allergies run comfortably through pollen-filled fields; and women skate effortlessly despite joint pain. Also, the types of drugs that are being made have started to change: more research dollars are being spent to develop lifestyle products, like those to combat impotency or hair loss. It's hard to argue the special place of an industry in society when it's hawking yet another potency product. Companies are starting to look like snake oil salesmen.

And it's not just the unseemliness of promoting lifestyle products that hurts pharmaceutical companies; some marketing practices have clearly evolved from aggressive to unethical. As a consequence, we don't trust pharmaceutical companies. Nor do we trust the doctors or scientists who work with them. Kenneth Rothman, an epidemiologist from Boston University, calls this "the new McCarthyism." Most people assume that investigators who have received research support from pharmaceutical companies cannot have an unbiased view. But where is the evidence for this in the vaccine-autism story? There exists not one example of a scientist or doctor serving on a vaccine advisory committee who, acting in his or her own financial interest, knowingly gave bad advice. In fact, after recommending vaccines for the nation's children, policymakers at the CDC—some of whom have performed studies funded by vaccine makers and are, therefore, closest to the data—invariably give these vaccines to their own children and grandchildren.

Although those who claim that vaccines cause autism have been quick to point out conflicts of interest among the scientists and doctors who disagree with them, few of the parent advocates, politicians, or scientists who speak against vaccines are without conflicts. Lyn Redwood, cofounder of Safe Minds, sued the federal government for compensation. So did Representative Dan

Burton's daughter, Danielle Burton-Sarkine. Both Redwood and Burton stood to financially benefit—either directly or indirectly— from the public's perception that vaccines cause autism. Robert F. Kennedy Jr. has a direct relationship with one of the largest product-liability law firms in the United States. Vijendra Singh, who testified at a Burton hearing that the MMR vaccine caused autoimmunity, received support from the Vaccine Autoimmunity Project. Richard Deth and Mady Hornig, both of whom claimed they had found, in their laboratories, how thimerosal caused autism, received funding from Safe Minds. And Andrew Wakefield received more than $800,000 from a personal-injury lawyer representing parents who were suing pharmaceutical companies.

So, if everyone appears to be in someone's pocket, who or what can be trusted? How can people best determine if the results of a scientific study are accurate? The answer is threefold: transparency of the funding source, internal consistency of the data, and reproducibility of the findings.

People have the right to know the funding source for scientific papers. For example, when Andrew Wakefield published his study of autistic children in the *Lancet*, he should have acknowledged that he had previously received money from Richard Barr and that Barr represented some of these children in a lawsuit against pharmaceutical companies. The irony in Andrew Wakefield's case was that not only did he fail to inform the *Lancet*'s readership of his funding source, but he failed to inform his co-investigators, most of whom later withdrew their names from his paper. Although funding sources should be reported in every scientific paper, they're probably the least important factor in judging a study's worth or reliability.

More important are the strength and internal consistency of the data. When Richard Horton found that Andrew Wakefield had received funds from a personal-injury lawyer, he was outraged. But Horton's anger should have been aimed at the obvious weaknesses in Wakefield's paper, not at his perceived motives. Andrew Wakefield had proposed that measles vaccine damaged children's intestines, allowing entrance of harmful toxins that caused autism. It was a hypothesis for which Wakefield offered not one shred of scientific evidence. Wakefield's paper shouldn't have been published not because he had received funds from a personal-injury lawyer but because his assertions were based on flimsy, poorly conceived science.

Probably the most important aspect of determining whether a scientific assertion is correct is the reproducibility of its findings. Superb, reproducible studies have been funded by pharmaceutical companies and poor, irreproducible studies have been funded independently, and vice versa. In the end, it doesn't matter who funds a scientific study. It could be funded by pharmaceutical companies, the federal government, personal-injury lawyers, parent advocacy groups, or religious organizations. Good science will be reproduced by other investigators; bad science won't.

Although the story of Andrew Wakefield and the MMR vaccine is an excellent example of the importance of reproducibility in assuring the validity of a scientific study, no story is more dramatic or more instructive than one that began at 1:00 p.m. on March 23, 1989. That's when Stanley Pons and Martin Fleischmann, nuclear physicists working at the University of Utah, announced they had caused nuclear fusion in a test tube. Pons and Fleischmann claimed they had taken a palladium electrode, inserted it into heavy water (deuterium), and observed a fusion event (when two lighter nuclei fuse to form a larger nucleus, releasing energy). This was big news. Pons and Fleischmann had found a way to provide safe, inexpensive, limitless energy. The media ate it up. Jerry Bishop, a superb science reporter, wrote about it in a front-page story in the *Wall Street Journal*. Utah legislators were so proud the breakthrough had occurred at the University of Utah that they allocated more than $4 million to establish the National Cold Fusion Institute on the university campus. Most scientists, however, were immediately skeptical, and for good reason: the Pons-Fleischmann experiment violated the first law of thermodynamics, which states that one can't get more energy out of something than is put into it. Later, when seventy different groups of physicists failed to find what Pons and Fleischmann had found, the promise of cold fusion disappeared. The building that housed the National Cold Fusion Institute now stands as a literal monument to irreproducible science.

· · · ·

OTHER ASPECTS OF OUR CULTURE ALSO DETERMINE HOW PEOPLE process scientific information. During the past few decades, doctors have started to treat patients differently. No longer do they always take a paternalistic, I-know-what's-best-for-you-so-don't-worry approach. Doctors are more apt to encourage patients to actively participate in their own medical care. And nothing has empowered people more than the Internet. Now patients have ready access to a wealth of information about health, medicine, and science. During a recent segment on the *Oprah Winfrey Show*, a celebrity mother was asked where she had gotten her medical information. "I attended the University of Google," she replied. J. A. Muir Gray, a British researcher and author of *The Resourceful Patient*, celebrates the culture of shared expertise. "In the modern world," he said, "medicine was based on knowledge from sources from which the public was excluded—scientific journals, books, journal clubs, conferences, and libraries. Clinicians had more knowledge than patients mainly because patients were denied access to knowledge. The World Wide Web, the dominant medium of the post-modern world, has blown away the doors and walls of the locked library." When Lyn Redwood and Sallie Bernard searched the medical literature for clues to the causes of autism, they were doing only what many doctors encourage parents to do: participate in the care of their children.

But empowering parents to make medical decisions comes with a price. Information on the Internet is typically unfiltered—anyone can say anything, and health advice can be terribly misleading. The vaccine-autism controversy is a good example. Doctors now constantly encounter parents who don't want to give their children MMR or thimerosal-containing influenza vaccines, fearing they might cause autism. "I've done my research," parents will say, "and I don't want my child to have that shot." By "research," the parents usually mean that they have perused a variety of Web sites on the Internet. But that's not research. If parents want to do genuine research on the subject of vaccines, they should read the original studies of measles, mumps, and rubella vaccines; compare them with studies of the combined MMR vaccine; and analyze the ten epidemiological studies that examined whether MMR caused autism. If they want to research thimerosal, they should read the hundred or so studies on mercury toxicity, as well as the eight epidemiological studies that examined whether thimerosal caused harm. This would take a lot of time. And few parents have the background in statistics, virology, toxicology, immunology, pathogenesis, molecular biology, and epidemiology required to understand these studies. Instead, they read other people's opinions about them on the Internet. Parents can't be blamed for not reading the original studies; doctors don't read most of them either. And frankly, few doctors have the expertise necessary to fully understand them, so they rely on experts who collectively have that expertise.

The experts who are responsible for making vaccine recommendations in the United States, and for determining whether vaccines are safe, serve the CDC, the AAP, the American Academy of Family Physicians, and the National Vaccine Program Office. And they do a pretty good job. During the past century, vaccines have helped to increase the life span of Americans by thirty years, and they have a remarkable record of safety. But if you're looking for a quote guaranteed to anger the American public, you need look no further than one delivered by Congressman Henry Waxman during Dan Burton's hearings. "Let us let the scientists explore where the real truth may be," said Waxman. In other words, let the experts figure it out.

Waxman's plea doesn't have much traction in today's society. Because of the Internet, everyone is an expert (or no one is). As a consequence, for some, there are no truths, only different experiences and different ways of looking at things. "This is the way that the world is going," laments Richard Smith, editor of the prestigious *British Medical Journal,* in an article titled "The Discomfort of Patient Power." "It's called post-modernism. There is no 'truth' defined by experts. Rather, there are many opinions based on very different views and theories of the world. Doctors, governments, and even the *British Medical Journal* might hanker after a world where their view is dominant. But that world is disappearing fast."

If doctors are going to encourage patients to make their own choices, they have to be willing to stand back and watch them make bad ones. They can't have it both ways. "Patients will often choose to ignore their doctors' advice and do something that their doctors regard as odd, even crazy," writes Richard Smith. Michael Fitzpatrick, the author of *MMR and Autism,* also sees danger in a culture in which experts cede their expertise. "We need to establish the foundations of an informal contract between parents and professionals that respects both our different spheres of expertise and—most importantly—the distinctions between them. Doing the best for our children means concentrating on being parents and leaving science to the scientists, medicine to the doctors, and education to the teachers." Fitzpatrick realizes that his request flies in the face of modern parenting. "So influential has the rhetoric of anti-paternalism become," says Fitzpatrick, "that this now appears a hopelessly old-fashioned proposal. But it is both principled and pragmatic. If I am having trouble with my car, I do not take to the Internet to study motor engineering; I take it to the garage and ask a mechanic to repair it. Even though I do not understand his explanation of the problem, I trust him. In a similar way, we put our trust in numerous people we encounter in our everyday lives. If we did not, society would simply collapse. The peculiarity of our current predicament is the selective withdrawal of trust from scientific and medical professionals, which is both unjustified and mutually damaging."

For many parents, the advice given by health care profession-als about vaccines is just one more opinion in a sea of opinions offered on the Internet.

． ． ． ．

WHEN MARK AND DAVID GEIER AND DEFEAT AUTISM NOW pro-posed to treat autistic children with mercury chelation, Lupron, restricted diets, and antibiotics—therapies not supported by any rigorous scientific studies—they were appealing to a long-standing, prevalent aspect of our culture: the lure of alternative medicine. As science reveals more and more about the workings of nature, this attraction hasn't weakened. If anything, it's got-ten stronger. More than 60 million Americans use supplements, megavitamins, herbs, and other alternative therapies in what has become a $40-billion-a-year industry.

America's commitment to alternative medicines is so strong that it led to the creation of a branch within NIH. In 1991, Con-gress passed a bill to create the Office of Alternative Medicine. Seven years later, this office became the National Center for Com-plementary and Alternative Medicine (NCCAM). Tom Harkin, a popular senator from Iowa, promoted the legislation. Harkin had been influenced by fellow Iowan Berkeley Bedell, who was convinced that his Lyme disease had been cured by eating spe-cial whey from Lyme-infected cows. Bedell's wasn't the only anecdote that convinced Harkin to carve out a special place on the NIH agenda. He, too, had gone against the advice of his doctors with amazing results: his allergies had been virtually eliminated by eating bee pollen. The center that now spends mil-lions of dollars to study alternative medicines was launched by these two experiences.

Although medicines that are alternative in today's culture may not be embraced by traditional, Western-trained doctors, that doesn't mean they don't work; it might mean only that they haven't been tested yet. But what worried many scientists and physicians about NCCAM was that alternative medicines would be exempt from the scientific method. Fortunately, that hasn't happened.

. . . .

ALTHOUGH IT'S BEEN AROUND FOR ABOUT 500 YEARS, THE SCIEN-
tific method is foreign to many. That's because most people don't understand what science is and what it isn't. People think of science as a body of knowledge or scientific societies or scientists. But it's really just a way of thinking about a problem. Indeed, most of us use the scientific method during routine activities. For example, if a radio doesn't work, we formulate a series of hypotheses: it isn't plugged in; the battery is dead; it isn't tuned to a local station. Then we go about testing each of these hypotheses separately until we have an answer. This is the scientific method—isolating one variable at a time and testing it.

. . . .

USING THE SCIENTIFIC METHOD, RESEARCHERS FUNDED BY NCCAM have now tested several alternative medicines. They have found that glucosamine and chondroitin sulfate don't treat arthritis; saw palmetto doesn't treat enlarged prostates; St. John's wort doesn't treat depression; shark cartilage doesn't treat cancer (based on the false belief that sharks don't get cancer); and Laetrile doesn't treat leukemia. Worse still, these studies have revealed a frightening aspect of alternative medicines: they can be quite dangerous. One natural alternative called compound Q, derived from the Chinese cucumber, was used to treat AIDS patients desperate for a cure. After it was found to cause severe toxic reactions and coma, compound Q was abandoned.

The lure of alternative medicines is understandable. When doctors fail to offer a cause or a cure for a particular disease, purveyors of alternative medicines often step into the void. Jerome Groopman, a professor of medicine at Harvard Medical School and the author of *How Doctors Think*, suffered severe, unrelenting pain in his wrist. "When I was a patient with a serious problem of uncertain outcome, I felt the powerful temptation to seek a magical solution," said Groopman. "Most doctors are sympathetic to this sensibility. But a good doctor distinguishes magic from medicine." For some, however, science is only an intrusion into beliefs that are as strong as religious convictions. Even when a particular notion is consistently refuted by scientific studies, they refuse to abandon it. During one of Dan Burton's hearings, a clinician named Kathy Pratt, who took care of autistic patients, was convinced that vaccines were the culprit "regardless of what the research tells us." Because science is the only discipline that enables one to distinguish myth from fact, Pratt's statement was particularly unsettling. "Uncovering [the laws of nature] should be the highest goal of a civilized society," says physicist Robert Park. "Not because scientists have a greater claim to a greater intellect or virtue, but because the scientific method transcends the flaws of individual scientists. Science is the only way we have of separating truth from ideology or fraud or mere foolishness." And science is enormously open-minded. If people believe they have a treatment for a particular disease or that one thing causes another, the scientific method can determine whether they are right. Suspected causes will be found to be true or not, and therapies will be found to work or not. "Things that are wrong are ultimately set aside and things that are right gain traction," said Stephen Strauss, former director of NCCAM. Strauss had a framed quotation on the wall of his office: "The plural of anecdotes is not evidence."

Although science is open-minded, the scientific method isn't terribly politically correct. To determine whether a medicine works, scientists establish a hypothesis, formulate burdens of proof, and subject those burdens to statistical analysis. Over time, a truth emerges. Something is either true or it isn't. And although our instinct is to be open to a wide range of attitudes and beliefs, there comes a time when it becomes clear that certain beliefs just don't hold up. MMR and thimerosal don't cause autism, and secretin, chelation therapy, and Lupron don't cure it.

• • • •

ALTHOUGH THE SCIENTIFIC METHOD HAS ALMOST SINGLE-handedly brought us out of the Dark Ages and into the Age of Enlightenment, it can be difficult to explain how it works. Here's the problem. In determining whether, for example, MMR causes autism, investigators form a hypothesis. The hypothesis is always formed in the negative, known as the null hypothesis. In the MMR-causes-autism case, the hypothesis would be, "MMR does not cause autism." Epidemiological studies have two possible outcomes: (1) Investigators might generate data that *reject* the null hypothesis. Rejection would mean that the risk of autism was found to be significantly greater in children who received MMR than in those who didn't. (2) Investigators might generate data that *do not reject* the null hypothesis. In this case, the risk of autism would have been found to be statistically indistinguishable in children who did or didn't receive MMR. But there is one thing those who use the scientific method cannot do; they cannot *accept* the null hypothesis. In other words, scientists can never say never. This means that scientists can't prove MMR doesn't cause autism in absolute terms because the scientific method allows them to say it only at a certain level of statistical confidence.

An example of the problem with not being able to accept the null hypothesis can be found in an experiment some children might have tried after watching the television show *Superman*. Suppose a little boy believed that if he stood in his backyard and held his arms in front of him (using Superman's interlocking thumb grip), he could fly. He could try once or twice or a thousand times. But at no time would he ever be able to prove with absolute certainty that he couldn't fly. The more times he tried and failed, the more unlikely it would be that he would ever fly. But even if he tried to fly a billion times, he wouldn't have disproved his contention; he would only have made it all the more unlikely. When scientists try to explain to the public the results of their studies, they always have this limitation in the back of their minds. They know the scientific method does not allow them to say, "MMR doesn't cause autism." So they say something like, "All of the evidence to date doesn't support the hypothesis that MMR causes autism." But to parents who are more concerned about autism (which they see and read about) than measles (which occurs uncommonly in the United States), this equivocation is hardly reassuring.

Another example of how scientists, respectful of the limits of the scientific method, fail to reassure the public can be found in a 2001 report from the Institute of Medicine (IOM) on the MMR vaccine and autism. This report, written after several excellent studies showed no relationship between the vaccine and the disorder, stated, "The committee notes that its conclusion does not exclude the possibility that MMR vaccine could contribute to autistic spectrum disorder in a small number of children." Those who wrote this report failed to point out that no study could ever prove MMR didn't cause autism in a small number of children because the scientific method would never allow it. But parents saw a door left open, and it scared them. Dan Burton picked up on this statement in one of his tirades against the IOM: "You put out a report to the people of this country saying that [the MMR vaccine] doesn't cause autism and then you've got an out in the back of the thing," screamed Burton. "You can't tell me under oath that there is no causal link, because you just don't know, do you?"

· · · ·

ANOTHER CHALLENGE FOR THOSE COMMUNICATING SCIENCE TO the public is explaining the difference between coincidence and causality. Because we're always looking for reasons for why things happen, this isn't easy.

When Andrew Wakefield reported the stories of eight children with autism whose parents first noticed problems within one month of their children's receiving MMR, he was observing something that statistically had to happen. At the time, 90 percent of children in the United Kingdom were getting the vaccine, and one of every 2,000 was diagnosed with autism. Because MMR is given soon after a child's first birthday, when children first acquire language and communication skills, it was a statistical certainty that some children who got MMR would soon be diagnosed with autism. In fact, it would have been remarkable if that hadn't happened. But parents of autistic children perceived their children were fine, got the MMR vaccine, and weren't fine anymore. (Although most children with autism show problems very early in life, about 20 percent will develop normally and then regress. It was this regression during the second year of life that caused some parents to blame MMR.) "Humans evolved the ability to seek and find connections between things and events in the environment," says Michael Shermer, author of *Why People Believe Weird Things*. "Those who made the best connections left behind the most offspring. We are their descendents. The problem is that causal thinking is not infallible. We make connections whether they are there or not." Physicist Robert Park agrees. "In humans, the ability to discern patterns is astonishingly general," he said. "Indeed, we are driven to seek patterns in everything our senses respond to. So far, we are better at it than the most powerful computer, and we derive enormous pleasure from it. So intent are we on finding patterns, however, that we often insist on seeing them even when they aren't there, like constructing shapes from Rorschach blots. The same brain that recognizes that tides are linked to phases of the moon may associate positions of the stars with impending famine or victory in battle."

For many parents, the association in time between their children's receipt of vaccines and the appearance of autism is far more convincing than epidemiological studies. That's because anecdotal experiences can be enormously powerful. Here's another example. A pediatrician in suburban Philadelphia was preparing a vaccine for a four-month-old girl. While she was drawing the vaccine into the syringe, the child had a seizure lasting several minutes. But imagine what the mother would

have thought if the pediatrician had given the vaccine five minutes earlier. No amount of statistical data showing that the risk
of seizures was the same in vaccinated or unvaccinated children
would have ever convinced her that the vaccine hadn't caused the
seizure. People are far more likely to be swayed by a personal,
emotional experience than by the results of large epidemiological
studies. "Popular induction depends upon the emotional impact
of the instances," said philosopher Bertrand Russell, "not on
their number."

Several years ago, a stand-up comedian, imitating a television
commercial advertising a book about the occult, showed how
hard it can be to distinguish cause from coincidence. Deepening
his voice, he said, "A woman in California burns her hand on a
stove. Her mother, three thousand miles away, feels pain in the
same hand at the same time. Coincidence?" Here he paused for
several seconds. "Yes!" he yelled, exasperated. "That's what
coincidence is!"

. . . .

ANOTHER ASPECT OF THE CURRENT CULTURE THAT MAKES IT DIF-
ficult to communicate science is the astonishing prevalence of
beliefs rooted in medieval times. "Two hundred years ago educated people imagined that the greatest contribution of science
would be to free the world from superstition and humbug,"
wrote Robert Park. "It has not happened. Ancient beliefs in de-
mons and magic still sweep across the modern landscape." According to a Gallup poll conducted in 1991, the statistics are
grim. About 50 percent of Americans believe in astrology, 46
percent in extrasensory perception, 19 percent in witches, 22
percent in aliens who have already landed on earth, 33 percent
in the lost continent of Atlantis, 41 percent in the notion that
dinosaurs and humans lived on earth at the same time (movies
haven't helped with this one), 42 percent in communication with
the dead, and 35 percent in ghosts. Thousands of people still
flock to Delphi, Greece, every year to gain energy from a place

they consider to be the center of the earth (based on the ancient belief that the earth was flat and could therefore have a center on its surface).

This is the lay of the land for scientists trying to explain cause and effect to the public.

. . . .

YET ANOTHER, MORE SUBTLE, ASPECT OF OUR CULTURE APPEARS throughout the vaccine-autism controversy. Two years after he published his paper in the *Lancet* claiming that MMR caused autism, Andrew Wakefield published "Measles, Mumps, Rubella Vaccine: Through a Glass Darkly." The phrase "through a glass darkly" is taken from Saint Paul's letter to the Corinthians (1 Corinthians 13:12) and refers to man's imperfect perception of reality. Wakefield's implication was that science—in this case the science that had claimed MMR was safe before licensure— couldn't be relied upon to get it right. And Wakefield believed that scientific studies that continued to absolve MMR couldn't be trusted to get it right either. Wakefield's capacity to set aside the studies that disproved his theory was based on a belief as powerful as a religious conviction. "He's very much like my father," said Wakefield's mother, Bridget. "If he believed in something, he would have gone to the ends of the earth to go on believing." When Andrew Wakefield first left England, he landed in Melbourne, Florida, with the Good News Doctor Foundation, whose logo features a stethoscope sitting on top of a Bible. The foundation describes itself as "a Christian ministry that provides hope and information on how to eat better and feel better, and minister more effectively as a result of a biblically based, healthy lifestyle." For Andrew Wakefield, the question of whether MMR caused autism had moved into the realm of faith.

While Andrew Wakefield continues to make religious references as he exhorts listeners to believe his theories, the most prominent religious figure in the vaccine-autism controversy is Lisa Sykes, an associate pastor at the Welborne United Method-

Protesters hold a sign in support of Andrew Wakefield during a hearing
before the General Medical Council on charges of misconduct. "Suffer little
children unto us" paraphrases Matthew 19:14, in which Jesus rebukes his
disciples for turning away a group of children. GlaxoSmithKline, Merck, and
Aventis Pasteur are the pharmaceutical companies that manufacture MMR
vaccine for children in the United Kingdom (courtesy of Getty Images).

ist Church in Richmond, Virginia. Sykes, who believes her son's autism was caused by thimerosal in vaccines, often delivers fiery speeches denouncing scientists at the CDC, FDA, and IOM, calling them "modern day deceivers." In April 2006, during an anti-vaccine demonstration in Washington, D.C., Sykes led those gathered in prayer "for the greedy and those who love power so much that they would seek profit over safety, and sacrifice children instead of wealth. We pray for those who have surrendered the truth, and government officials who have failed to seek it. These, too, like so many of our injured children, cannot see, they cannot hear, and they remain silent." In February 2007, the United Methodist Virginia Conference published its Lenten Devotional, in which Sykes interpreted scripture and issued a call to action: "My son is disabled," she said, "unnecessarily injured by mercury he received in vaccines. Like Abram, we are cast down. The era of administered mercury is the darkest part of the night."

"I think about symbols," said Kathleen Seidel, in reference to cleansing the autistic child's body of mercury. "And there are a lot of powerful symbols that are part of this whole hysteria, the whole concern over vaccines: symbols of purity and defilement and of sin and redemption." One of the Rescue Angels of Generation Rescue (the organization dedicated to mercury chelation) proclaimed that with chelation, "We're helping [a child's] body do what God intended it to do." Where science and medicine have failed to find a cause or cure for autism, some have put their trust in the certainty, absolutism, and occasional zealotry of Andrew Wakefield, Lisa Sykes, and Mark Geier, people who ask their followers to have unquestioning faith in theories contradicted by scientific evidence.

· · · ·

ANOTHER ASPECT OF OUR CULTURE—AND ONE REASON THE MMR and thimerosal controversies gained immediate attention—is that it's easy to scare people. For example, beginning in the 1960s and 1970s, rumors that people had put razor blades into apples or poisoned Halloween candy swept across the nation. Everyone believed it. As a consequence, parents insisted that their children eat only prepackaged candy, schools opened their doors so that trick-or-treaters could have a safe environment, and hospitals offered to X-ray candy bags. In their book, *Made to Stick: Why Some Ideas Survive and Others Die*, Chip and Dan Heath examined the widespread belief that trick-or-treaters were at risk. They found that since 1958, no one had ever been harmed by a stranger's Halloween candy. The urban myth had been spawned by two events. First, a five-year-old boy had overdosed on his uncle's heroin; to cover his tracks, the uncle put heroin on the child's candy. Second, a father, in a twisted attempt to collect insurance money, killed his son by sprinkling cyanide on his candy. "In other words," wrote the Heaths, "the best social science evidence reveals that taking candy from a stranger is perfectly okay. It's your family you should worry about."

Although the fear of tainted Halloween candy isn't based on a single occurrence, it hasn't died, and it probably never will. Both California and New Jersey have passed laws specifically designed to punish candy tamperers. Similarly, laws banning thimerosal-containing vaccines have passed in several states despite clear evidence that these vaccines aren't harmful. It's much easier to scare people than to unscare them.

· · · ·

A FINAL CULTURAL ASPECT—AND YET ANOTHER REASON THAT the mercury-in-vaccines controversy stuck—is that it's easy to appeal to the notion that we live in a sea of poisonous metals, toxic chemicals, and environmental pollutants. To be sure, some toxins in the environment can be quite dangerous. In the United States, high levels of lead in paint caused severe neurological problems in many children. And in Japan, the Minamata Bay disaster showed just how devastating large quantities of mercury can be. But these aren't typical stories. For example, the media declared that dioxin, the chemical buried under the Love Canal in upstate New York, caused birth defects and miscarriages; that hexavalent chromium, the chemical used by Pacific Gas and Electric to coat its pipes (and the subject of the movie *Erin Brockovich*), caused a variety of illnesses from nosebleeds to cancer; that trichloroethylene, the chemical dumped by the W. R. Grace tannery into the local water supply (and the subject of the book and movie *A Civil Action*), caused a cluster of cancer cases in Woburn, Massachusetts; and that Alar, a pesticide featured on a *60 Minutes* program titled "A is for Apple," caused cancer. None of these stories was supported by subsequent scientific studies. But studies showing that certain chemicals in the environment aren't harmful are far less compelling than personal testaments, riveting television shows, and blockbuster movies claiming that they are. Steven Milloy, a graduate of the Johns Hopkins School of Hygiene and Public Health, the author of *Junk Science Judo*, and the creator of the popular Web site JunkScience.com, laments how the media are attracted to stories that scare people but not to those that reassure them. When Milloy approached *Dateline NBC* with a story about how fears of small quantities of dioxin were unfounded, he was rebuffed. "I was interviewed about our [dioxin] study by seemingly interested staff of the television news magazine *Dateline NBC*," recalled Milloy. "After about twenty minutes of questions, it finally dawned on the staff person. 'So, this isn't a scare story?' she said. 'Then my producer won't be interested.'"

Recently, the comedy team of Penn and Teller filmed a three-minute video for YouTube that showed just how easy it is to appeal to the public's concern about chemicals in the environment. They sent a friend to a state fair to collect signatures on a petition to ban dihydroxymonoxide. Dihydroxy (two hydrogen atoms) monoxide (one oxygen atom) is H_2O—water. The petitioner never lied. She said that dihydroxymonoxide was in our lakes and streams, and now it was in our sweat and urine and tears. We have to put a stop to this, she urged. Enough is enough. By using its chemical name, she was able to collect hundreds of signatures to ban water from the face of the earth.

The media bias toward stories that scare rather than reassure has left the public with a poor understanding of risk. "Hundreds of thousands of deaths a year from smoking is old hat," writes Michael Fumento in *Science Under Siege*, "but possible death by toxic waste, now that's exciting. The problem is [that] such presentations distort the ability of viewers to engage in accurate risk assessment. The average viewer who watches story after story on the latest alleged environmental terror can hardly be blamed for coming to the conclusion that cigarettes are a small problem compared with the hazards of parts per quadrillion of dioxin in the air, or for concluding that the drinking of alcohol, a known cause of low birth weight and cancer, is a small problem compared with the possibility of eating quantities of Alar almost too small to measure. This in turn results in pressure on the bureaucrats and politicians to wage war against tiny non-existent threats. The 'war' gets more coverage as these politicians and bureaucrats thunder that the planet could not possibly survive without their intervention, and the vicious cycle goes on." As a consequence, people are more frightened by things that are less likely to hurt them. They are scared of pandemic flu but not epidemic flu (which kills more than 30,000 people a year in the United States); of botulism, tsunamis, and plagues but not strokes and heart attacks; of radon and dioxin but not French fries; of flying but not driving; of sharks but not deer; of MMR but not measles; and of thimerosal-containing influenza vaccine but not influenza. During the Alar scare, one mother sent state troopers after a school bus to confiscate an apple she had put in her child's lunch bag; another called the International Apple Institute to ask if it was safe to pour apple juice down her kitchen drain or if she should take it to a toxic dump site.

• • • •

THE VACCINE-AUTISM CONTROVERSY HAS SHOWN JUST HOW DIFficult it can be to communicate science to the public. Fortunately, during the past few years, many studies have investigated the true causes of autism; ironically, the media's constant focus on vaccines has made it difficult for the public to hear about them.

CHAPTER 11

A Place for Autism

When you begin to touch your heart or let your heart be touched, you discover that it's bottomless.

—Pema Chodron

The first clue to the cause of autism is that it's genetic. (Genes contained in chromosomes in the nuclei of cells provide a blueprint for cell function.) Researchers have shown that autism is genetic by studying twins. They found that when one identical twin had autism spectrum disorder, the risk to the second twin was greater than 90 percent; in contrast, when one fraternal twin had autism, the risk to the second twin was less than 10 percent. Because identical twins share the same genes and fraternal twins don't, these studies proved that autism was in large part genetic.

When researchers first described the genetic basis of autism, they hoped one gene was responsible. It wasn't to be. The genetics of autism is far more complex. Edwin Cook, a child psychiatrist and autism researcher at the University of Illinois Medical Center in Chicago, describes autism as "the most heritable of the neurodevelopmental disorders that are complex in origin." So there was good news and bad news. The good news was that the cause of autism could be found in the genes (meaning researchers could determine the function of these genes); the bad news was that autism was associated with many genes (meaning it wasn't going to be easy to do this). Later, researchers found that although all autism was genetic, not all autism was inherited. Some parents with no family history of autism had children with the disorder. Apparently, for some children, spontaneous mutations occurred in certain genes. In December 2007, researchers in Canada further confirmed the genetics of autism.

More clues followed. Researchers tried to determine *when* these particular genes started to malfunction. To do this, they used a clue that many parents had lying around the house: home movies. After looking at movies of children's first birthday parties, researchers could predict which children would later be diagnosed with autism and which wouldn't. (This was exactly what Eric Fombonne had done when he examined videotapes of Michelle Cedillo's first birthday party during the Omnibus Autism Proceeding.) Some children with autism could be detected even earlier. While looking at home movies taken before a child's first birthday, Canadian researchers found abnormal behaviors in some children as early as six months of age.

Then researchers made an even more surprising discovery: some babies started on the road to autism *before* they were born. Studies of thalidomide, a sedative that had never been licensed in the United States but had been widely used in Europe and Canada, were particularly interesting. Researchers found that mothers who took thalidomide—known to cause severe birth defects, including shortened arms and legs—delivered babies with an increased risk of autism. These studies showed that factors in the environment (in this case, a drug) were capable of causing autism.

Then researchers found it wasn't only drugs that could cause autism; viruses could do it too. If mothers were infected with rubella virus (German measles) early in pregnancy, their babies were at higher risk of autism. More than any other clues, thalidomide and rubella showed that environmental factors could influence the development of autism.

These findings led to the next obvious question: if thalidomide and rubella virus could cause autism early in pregnancy,
wasn't it at least plausible that the mercury in thimerosal could
do it too? To answer this question, Judith Miles, a developmental specialist working at the Thompson Center for Autism and
Neurodevelopmental Disorders at the University of Missouri,
studied pregnant women who had received thimerosal in a prep-
aration called RhoGam. RhoGam is given during pregnancy
when one particular protein, called the Rh factor, is present on
the unborn baby's red blood cells but not on the mother's. Miles
found that thimerosal given to pregnant women didn't cause
autism in their babies. Miles's findings were consistent with
those on the massive mercury poisonings that had occurred in
Minamata Bay and Iraq, where babies were not at greater risk
for autism.

Another interesting breakthrough came in March 2007, when
the Autism Genome Project, a collection of research groups from
more than fifty institutions, unveiled the results of a five-year
study. Researchers in the project who were exploring cells of
autistic children had found abnormal proteins in the area between nerve cells, called the synapse. Two specific proteins,
neurexin and neuroligin, appeared to be involved. One of the
lead researchers, Thomas Bourgeron from the Pasteur Institute
in Paris, was the first to propose that synapses in the brain were
critical to the development of autism. "People in the field are really accepting that this is a pathway which is associated with
autism," said Bourgeron. "When we published the neuroligin
[report in 2003], nobody believed it."

Interactions between one brain cell and another at the synapse didn't explain everything, however. Autistic children who
had a genetic defect only in synaptic proteins were rare, so other
genes were clearly involved. In the end, it's unlikely that even a
handful of genes will explain everything. That's because the
symptoms and timing of autism are so varied and diverse. Some
children have very mild symptoms, others quite severe. Some

children develop symptoms slowly over the first few years of life, others appear to develop normally and then regress. But recognizing the different expressions of autism and the likely variety of genes associated with them is an important and exciting start.

As autism genes have become better defined, people have become more and more excited about the possibility of specific treatments. But excitement should be tempered by experiences with two other genetic disorders: sickle cell disease and cystic fibrosis. Sickle cell disease is caused by a single change on a single gene. No genetic disorder is simpler. That one change results in the production of abnormal hemoglobin, causing red blood cells to collapse. (Hemoglobin is the protein in red blood cells responsible for carrying oxygen to the body.) Although children with sickle cell disease have benefited from the discovery of antibiotics and from advances in blood transfusion, knowing the exact single genetic change that causes the problem hasn't improved the lives of its sufferers. Similarly, cystic fibrosis, a disease that causes progressive loss of lung function, is caused by a single gene. Unlike sickle cell disease, the cystic fibrosis gene contains hundreds of changes, but it's still just one gene. And similar to sickle cell disease, the abnormal gene that causes cystic fibrosis makes just one protein. Again, this knowledge hasn't led to specific treatments for cystic fibrosis. The genetics of sickle cell disease and cystic fibrosis are relatively simple. One abnormal gene makes one abnormal protein that causes disease. For autism, the story is far more complex. It's not just that there might be hundreds of different changes on one gene; there might be hundreds of different changes on many genes. For these reasons, even after the genetics of autism has been clearly defined, treatments might not be just around the corner.

• • • •

MANY PARENTS INSTINCTIVELY UNDERSTAND THAT THERE IS NO immediate cure for autism and that vaccines are not the cause.

During the vaccine-autism controversy, four parents of autistic children, sickened by the label of "vaccine-damaged," bravely entered the fray. All have been a wonderful resource for parents trying to understand their children's disabilities.

Camille Clark, known to thousands of bloggers as Autism Diva, is a remarkable woman. Raised in Davis, California, Clark recognized that, like her daughter, she too was mildly autistic. "I knew I was very different," she remembered, "and was always looking for clues for why that was. In situations that were overwhelming you could find me in a corner reading a book. I learned to read and write early, and I really liked to read. I remember feeling awkward in talking to people I didn't know. I still feel this way."

In the 1970s, Clark moved to Idaho, where she met her future husband. They had two children. "We were two very odd people, coming from families populated by some very odd people," remembered Clark. "Somehow we produced one very odd child and one normal child." But the marriage didn't last, and Clark moved back to Davis to be near her mother.

By 2000, Clark, who was then cleaning houses for a living, started taking courses at a local community college, later graduating from the University of California at Davis with a degree in psychology. At the same time, she was trying to figure out what was happening to her daughter. "I had been searching for several years trying to figure [it] out," she remembered. "Though the signs of autistic spectrum disorder were there from very early, no one ever said, 'Your child has poor eye contact and odd, restricted interests. Have you noticed all that flapping? That's autism.' Of course, I had noticed all that flapping, but it never occurred to me that it was a symptom of something. Sometime in 2000, I was on the Internet and found a description of hyperlexia, the very early ability to read. I hadn't heard of it before. The descriptions of hyperlexia frequently referred to Asperger's Syndrome and autism. Suddenly, I realized that this was a description of me. I found information about Internet groups for adults with Asperger's, and I joined a couple of them."

Clark was drawn to parents' personal stories but appalled by their lack of scientific rigor. "I joined Quackwatch's health-fraud listserv and started asking experts questions like, 'What is oxidative stress?' and 'Is this right what they are saying about mercury and chelation?' I found this whole world of science that was in direct opposition to the nonsense going on in the media about a generation of children destroyed by vaccines. I had written a few things about the anti-vaccine autism epidemic hysteria and posted them to the autistic advocacy group and to the health-fraud listserv, when someone on that list suggested that members start blogs. That's how I was introduced to blogging."

Clark would soon become the best known, most widely read, and arguably the most irreverent autism blogger. "On the autistic advocacy Internet group we had many discussions about how awful self-proclaimed 'experts' and professionals were and how badly they reacted to any kind of criticism or questioning," recalled Clark. "They had this How-dare-you-question-me! attitude. I thought of them as autism divas. So I created Autism Diva and decided that she'd be just as uppity as they were, and she'd tell them what was what, straight up, and not apologize for doing so. Over time, mostly because of this blog, I made more and more contacts with people who were scientists and skeptics, most of whom were also parents of autistic children or who were autistic themselves. These people taught me more about the details of why the quack therapies couldn't work the way they were hyped."

Clark has paid a price for her skepticism. "I have been harassed by phone and by e-mail," she says. "Among the people who hate me, most of them think that I have been paid to write about autism quackery by 'Big Pharma' or some other nefarious entity. But I haven't made any money from blogging, not a penny. It has cost me a little in cash, but mostly it took a lot of time. The drive behind Autism Diva is partially self-preservation and looking out for my own family. I don't want people to call my child 'poisoned' or an 'empty shell'; neither do I want to see vaccination rates fall so low that my family or anyone else's is impacted by outbreaks of vaccine-preventable diseases. I was devastated by the death of Abubaker Tariq Nadama because I had feared that chelation might kill a child."

Now a laboratory technician for an alternative energy company, Clark hopes someday to work in the field of autism research. Although Clark doesn't have a cure for autism, she has several suggestions for parents that she hopes will make their lives easier. First, she asks parents to pay attention to the unique skills of their autistic children. "The very wiring of an autistic brain means that the autistic person is likely to have significant abilities," she says. "Those abilities aren't always 'saleable'—such as a talent for mathematics might be—but autistics usually have very keen memories and almost always have a desire to collect facts. Autistic abilities shouldn't be seen as disabilities or freakish 'splinter skills' just because they are less common among non-autistic people." She also asks parents not to be disheartened by dire predictions from their doctors. "Professionals will be unable to tell you what your autistic child will be able to accomplish in the future," says Clark. "But the very same thing is true for non-autistic children. No one has a crystal ball. Don't let that fact make you despair." And she warns parents to be suspicious of dramatic cures. "All children have sudden developmental leaps and times with less apparent development," she says. "This bumpy trajectory may be more pronounced in autistic children [and] makes parents targets for quack therapies of all kinds because there will always be someone who can say, 'Before we put our child on this diet she never made eye contact, but now she does!'" Finally, Camille Clark offers a message of love. "If your child can't say the words 'I love you' now, or even ever, that doesn't mean your child doesn't feel love for you," she says. "A deaf child might never learn to speak the words 'I love you' and deaf parents may never hear the sound 'I love you.' But deaf parents don't assume that their children don't love them. If parents respond with love to their autistic children's non-standard ways of showing love, then parents and children will feel the bond that they both desire."

• • • •

PETER HOTEZ IS THE FATHER OF A SEVERELY AUTISTIC DAUGHTER. He has testified before congressional committees, participated in press conferences with public health officials, and often been quoted in newspaper and magazine articles. More than any other parent, Hotez has spoken out publicly against the notion that vaccines cause autism.

Hotez grew up in Hartford, Connecticut, dreaming of becoming a scientist. "In my early childhood years, believe it or not, I wanted to study tropical diseases," he recalled. Toward that end, after graduating from Yale University, Hotez entered the Rockefeller Institute in New York City for his doctorate. There, surrounded by Nobel Laureates, he began his lifelong study of hookworms, a common, devastating infection in the developing world. Later Hotez got his medical degree from Cornell University and completed his residency in pediatrics at Massachusetts General Hospital, part of Harvard Medical School. He is now the Walter G. Ross Professor and chairman of Microbiology and Tropical Medicine at George Washington University School of Medicine.

Peter and Ann Hotez have four children. The first two, Matthew and Emy, are students at George Washington; the third, Rachel, is now fifteen years old; the fourth, Daniel, is ten. "We noticed that something was wrong [with Rachel] early on," recalls Hotez, "but we couldn't put our finger on it. Ann would say that she wasn't as 'huggy' as the other kids. That she had much more of a shrill cry. And that she was just a really hard baby. We actually started taking her to pediatric neurologists because we felt something was wrong. But there was nothing really objective on her physical exam that they could pinpoint." Later, at the Yale Child Study Center, doctors diagnosed Rachel's autism.

Although he was confident Rachel would get the care she needed during the day, Hotez was overwhelmed by his daughter at home. "We wound up sending her to private schools," said Hotez. "But we'd still have her home in the summer, and we'd still have her outside of school hours. When you add up all the hours she had in school, it was less than 20 percent of the time. And the rest of the time she was at home making life horribly difficult. She would run away to people's houses. She was very rigid, and she wouldn't go in the car. So you'd have to stay at home. It got to the point where Ann and I literally did not go out to dinner for a decade. She was always destroying the house, lining things up, putting things in the toilet. [And she had] no sense of self hygiene." The impact of Rachel's difficulties wasn't limited to the demands of her care. "Because of Rachel we had to ignore the needs of the other kids. And it had a lot of financial implications as well, which is that Ann couldn't go back to work. [My solely] supporting a family was hard. As a result, we accumulated horrific debts that we still carry with us to this day. So we had terrible emotional and financial issues all around this one illness, which is one of the reasons I can understand why parents get so angry. Your first reaction is always that somebody has to be blamed for this."

Although Hotez was convinced by the studies exonerating vaccines, he was equally convinced by his experiences with his daughter. "We knew very early that there was something wrong," he said. "So there was never any question that it was a vaccine or something that we did after birth. Having two normal kids before Rachel, and with my being a pediatrician, we had our antennae out for any sort of problems. In Ann's case, she was always wondering what she could have done during her pregnancy to cause this."

With increasing anguish, Hotez watched the vaccine-autism controversy unfold. Mostly he was frustrated by the failure of public health officials to stand up for the science that put the vaccine question to rest. "One of the reasons I decided to speak out on this," said Hotez, "was that I was really appalled at the way that the U.S. Public Health Service has been silent on this issue. I mean, not a word. During all of these Omnibus hearings, did the surgeon general ever speak out on this issue? The CDC is barely vocal on it. They held one press conference that I was part of and then afterwards they didn't allow us to answer questions from the press. [The National Vaccine Program] Office can't really speak out on it. The secretary of health and human services says nothing. And people are looking to our public health leaders for some guidance on this. I think [the CDC] is still a respected office despite what some of the autism parent groups say about it. But there are just no public statements, no guidance. When you looked at some of the articles that were coming out in the newspapers, I really found them infuriating: every time they talked about autism, it was always in the context of this vaccine-autism debate, when in fact vaccines have absolutely nothing to do with autism. There is a need for more research into the genetics of autism, and there's a need for special services, but the whole debate is being skewed, partly because of the silence of our government health officials. I felt the need to speak out because somebody who didn't have a horse in the race needed to say something."

Hotez feels the constant focus on vaccines has hurt autistic children. "The hardest part about taking care of an autistic child is that you really are alone," he said. "There are no services that are out there to help you in the home. And that's where the kid is 80 percent of the time. To this day, one of the reasons that I believe that we are at least ten years behind in providing the right kind of services for autistic children is because of the distraction that this whole vaccine-autism debate has caused. It's led to a lack of focus on what's really needed. I get very angry at a lot of these autism groups, like Safe Minds. It's so difficult for me not to want to shake them and say 'Don't you realize that you're really doing a disservice to parents, not a service?' And they're so self-righteous. They don't speak for all autistic parents. They're certainly not speaking for me. Everyone is terrified of these parent groups. One of the advantages of having an autistic kid, one of the few, is that I don't have to be terrified. I've suffered as much as any parent has, and I can just tell it like it is."

Like Camille Clark, Hotez has paid a price for his outspokenness. "When I became a public spokesperson, it's been very hurtful to see some of the stuff that's on the blogs. They'll say that I'm paid off by industry. Most of the blogs that are about me actually portray me as what I like to call the modern-day equivalent of one of Vladimir Lenin's useful idiots. That I'm being duped by drug companies."

• • • •

KATHLEEN SEIDEL HAS ALMOST SINGLE-HANDEDLY EXPOSED THE unsavory allegiances of those who proffer cures for autism. She's also been a constant, unshakable thorn in the side of those who have hijacked discussions about the cause of autism.

After her child was diagnosed with Asperger's Syndrome, Seidel saw the world of autism around her. "I realized that I'd been interacting with [people] on the spectrum, and it didn't seem so threatening," she said. "The net is getting broader. People are getting diagnosed who didn't get diagnosed before. And I realized that there was this disjuncture between the medical prognoses that are offered in the literature that are based on more seriously handicapping conditions and [milder forms] of the spectrum." In time, Seidel came to appreciate her child not as damaged goods, but as part of a larger, neurologically diverse group. "The ['damaged-goods'] label is something that physically disabled people have protested," she said. "Like the Jerry Lewis telethon, patheticizing portrayals of children. I didn't like these catastrophic descriptions. There is no autism separate from a human being. Choosing to describe a child as soulless—that may be someone's grief stricken surface misimpression, [but] it was fundamentally offensive to me. There was a point where we realized that our kid is the same person as before the diagnosis."

In her search for advice and solace on the Internet, Seidel became progressively more angry at those who claimed to speak for her child. "When Generation Rescue did their brave new world start-up in the spring of 2005, I was just appalled by the statements that they were making, the broad generalizations. All autism is caused by mercury poisoning? Excuse me. Did it ever occur to you that someone might object to having their family members labeled as inherently toxic?" So Seidel joined the listserv that had been launched by anti-vaccine activists. "I signed onto the *Evidence of Harm* mailing list because I wanted to check out the conversations—to see the people who were promoting [David Kirby's] book and to see what was going on. I was blown away by the hostile judgments towards parents who did not believe in the whole mercury-in-vaccines business. [And] I was blown away by the overwhelming negativity and hostility that was expressed by so many people toward anyone who disagreed with their conclusions about scientific subjects." Seidel entered the fray by writing an open letter to David Kirby. "That's when I wrote my 'Evidence of Venom' letter," she recalled. "I thought

At a march on Capitol Hill, July 20, 2005, the notion that children with autism are poisoned is expressed in the sign, "Thimerosal in vaccines, ingredient of mass destruction, 100's of thousands of brain damaged children" (courtesy of Getty Images).

maybe this guy is just really naïve. Maybe he thinks that he's fighting the good fight. But does he realize that there are a whole lot of us out there who don't believe in this whole vaccine thing, for whom it has no relevance in our lives and our experience, who are kind of offended at the willingness of certain people to go out beating this drum saying that all autistics are poisoned? And I was offended by the virulence, the condemnation [by the anti-vaccine camp] of medical professionals."

Seidel, who describes herself as a liberal Democrat, suddenly found herself on the other side of the fence. "I felt a certain sympathy for those people who were being demonized, absolutely demonized, just by their choice to work in a certain area of medicine. You know, how just is this? It's ironic to me that this whole business with the vaccines and thimerosal and autism has brought me to positions that put me in line with pharmaceutical companies. But I find that I have common ground on some very specific things with people who have radically different political dispositions than I have."

The more she studied the subject, the more Seidel became certain that vaccines had nothing to do with autism, and she was angry at the doctors and scientists who had convinced parents into thinking they did. "I don't want to rag on parents," she said. "Parents are trying to help their kids. And they are trusting people who appeal to them for all sorts of reasons. But, boy, people who exploit families piss me off. When I read about a doctor who suggests to a parent that they take out a second mortgage on their house so that they can buy a home Hyperbaric Oxygen Therapy chamber, I want to cry. People talk about how it's economically devastating to take care of an autistic child. Well, it's certainly not helped by doctors who imply that their treatments may be able to reconfigure a child's brain when they use words like *cure* and *recovery*. I think *recovery* is a really loaded term, and parents have so much hope for their children."

Because Kathleen Seidel finds that autistic children display a wide range of symptoms, she is reluctant to offer advice; but she does offer a plea for patience. "People often come to doctors and educators expecting that they will have answers for them. But they don't. Doctors in particular, because they come from a scientific background, are accustomed to uncertainties. Some things we just don't know. Sometimes the best answer is there is no answer. One of the reasons that Mark Geier's testimony has been thrown out of court is that he doesn't acknowledge an alternate possibility, which is that the cause of autism is unknown."

• • • •

ROY RICHARD GRINKER GOT HIS DOCTORATE FROM HARVARD University and is now a professor of anthropology and director of the Institute for Ethnographic Research at George Washington University. He's also the father of an autistic teenage daughter, Isabel. Drawn into his daughter's world, Grinker recently wrote a book about her and about how autistic children are treated in different cultures. It's called *Unstrange Minds: Remapping the World of Autism.* Grinker is struck by how far we've come in our perception of autism. His message is one of hope. "Just thirty years ago," he wrote, "my daughter Isabel might have been labeled mentally retarded, and there would have been little opportunity for her to find her place in the world. Isabel would probably have been placed in a residential institution with a minimal education plan, where the symptoms of her autism would have worsened. A mildly autistic child living at the same time at home would have been teased and bullied mercilessly, would have had little access to special-education services, and would have failed at school and suffered profound emotional distress. Today, my teenage daughter is mainstreamed into a high school classroom."

Grinker is excited by the explosion in autism research. Recently, he received a grant from the National Alliance for Autism Research to conduct the first-ever epidemiological study of autism in Korea. Like Clark, Seidel, and Hotez, Grinker doesn't believe vaccines cause autism, but he sees a light at the end of the tunnel: "Federal, state and local agencies in many countries, especially the United States, Canada, and the United Kingdom, have been mobilized to manage the heavy public health burden of autism. Special-education programs are expanding; in the United States, new money is pouring out of the National Institutes of Health into autism research, and donors are contributing millions of dollars to parent advocacy organizations, private schools, and research foundations. Even scientists who never before had an interest in autism are joining an increasingly long parade of autism researchers." But Grinker worries that the vaccine issue has diverted both research funding and parents' attention. "In a remote Appalachian mountain community," he wrote, "I met impoverished parents of autistic children, moms and dads who never went to high school, can barely read or write, and certainly have no Internet access. But they all knew the word *thimerosal*."

· · · ·

DESPITE THEIR STRUGGLES, CAMILLE CLARK, PETER HOTEZ, AND Roy Richard Grinker have each found something to hold on to, something good, in their experiences with autistic children. Clark has found joy in looking at life through her daughter's eyes. "My autistic spectrum child has an unusually beautiful view of the world," she says, "one that is unsullied by cynicism, uncomplicated by a sense of competition, and unmarred by selfishness. This unusual person is unfailingly kind and never has a desire to meddle in other's lives. It wouldn't occur to [her] to lie to get some kind of personal advantage. I am fairly cynical and aware of how duplicitous people can be. [But] my daughter keeps me from becoming entirely bitter about humanity. Besides helping me to see the world differently, my child has helped me to see myself more accurately; having the world continually described to me in rosy terms shows me where my perspective is too bleak and unforgiving."

Peter Hotez found his daughter's autism led to a heartening family dynamic. His elder daughter, Emy, "really rallied around Rachel," said Hotez. "She got interested in special needs children, volunteering at [Rachel's] school. And [she's] even thinking of becoming a psychiatrist or a neurologist as a result. So it really had a profound influence on her for the better."

Roy Richard Grinker found strength in what he learned from his daughter. "Isabel has taught me that the unexpected, even the beautiful, can emerge even from the undesirable, like a lotus growing out of the mud, its beauty and purity unsullied by its origin," he wrote. "The beauty can be found in a single person, inside of whom there is something—no, not something 'normal,' but a brilliant light or an inner truth struggling to blossom. So, when people pity me for my daughter, I don't understand the sentiment. I work hard for Isabel, but I don't regret it or feel sorry for myself. At the end of the day when I tuck her in, she's not a case of autism, or even a child with social deficits and language delays. She's simply my daughter. My job is to clear the land for whatever growth is to come, even if, sometimes, no one else believes it will happen."

EPILOGUE: "NEXT, ON *OPRAH*"

So it goes.

—Kurt Vonnegut, *Slaughterhouse Five*

On September 18, 2007, Oprah Winfrey interviewed actress Jenny McCarthy on her nationally televised daytime show. Few women are more respected than Oprah Winfrey; her philanthropy, goodwill, and common sense have made her one of the most trusted and powerful people in America. McCarthy came to Oprah to talk about her new book, *Louder than Words*, the story of how she had cured her son's autism. For authors, an appearance on the *Oprah Winfrey Show* is as good as it gets; Oprah's Book Club has launched a generation of readers.

Oprah: So what do you think triggered the autism?

McCarthy: I do have a theory [based on] mommy instinct. You know everyone knows the stats with one in 150 children with autism. What I have to say is this. What number does it have to be? What number does it take for people just to start listening to what the mothers of children with autism have been saying for years—which is that we vaccinated our babies and something happened. That's it. And we don't know why. I told my pediatrician, "Something happened." And the reaction I got was of making me feel dumb and stupid, and I felt very alone in this.

Oprah: How were you made to feel alone?

McCarthy: Because right before my son got the MMR shot I said to the doctor, "I have a very bad feeling about this shot. This is the autism shot, isn't it?" And he said, "No! That is ridiculous. It is a mother's desperate attempt to blame something on autism." And he swore at me. And then the nurse gave [my son] that shot. And I remember going, "Oh, God, no!" And soon thereafter I noticed a change. The soul was gone from his eyes. And the thing is [that] I'm not against vaccines. We do need them. I am all for them. But there needs to be a safer vaccine. Something needs to be done, and people need to start listening to what the moms are saying.

Oprah: Of course, we talked to the Centers for Disease Control and asked them whether there was a link between autism and childhood vaccines. And here's what they said. "We simply don't know [Oprah paused for several seconds] what causes most cases of autism, but we're doing everything we can to find out. The vast majority of science to date does not support the association between thimerosal in vaccines and autism, but we are currently conducting additional studies to determine what role, if any, thimerosal in vaccines might play in the development of autism.

McCarthy: My science is Evan, and he's at home. That's
            my science. [Loud, thunderous applause].

The day after Oprah's interview with Jenny McCarthy, anxious parents flooded the CDC with calls about the safety of vaccines; pediatricians were also inundated with questions.

· · · ·

THE FOLLOWING WEEK, ON SEPTEMBER 24, 2007, MCCARTHY AP-
peared on *Good Morning America* with Diane Sawyer.

McCarthy: I wish I would have known what I know
            now, because I know things would have
            ended up a lot differently for Evan.

Sawyer: What do you mean?

McCarthy: With all of my belief and knowledge now in
            terms of vaccinations and toxins in the world
            and infections, I do believe like many, many
            thousands of moms out there, that it pushed
            them into what we're calling autism. And
            after doing my *Lorenzo's Oil* and going
            online and researching, I found that by fixing
            this, the sickness, the toxins, he was getting
            better. But the pediatricians aren't offering
            up this kind of information. It's all up to the
            moms.

McCarthy couldn't believe the gold mine she had found on
the Internet, reading about what caused autism and how to cure
it. "Evan was possibly born with a weaker immune system," she
wrote. "Getting vaccinated wreaked havoc in his body, and
mercury caused damage to his gut . . . and one could see the result of this damage when he consumed wheat or dairy. It messed
with his little body so much that he wouldn't respond when his
name was called; he behaved like a drunk, and the list goes on."

On the Internet, McCarthy learned that autism could be treated with special diets. "After beginning a gluten- and casein-free diet with vitamin supplements, mothers reported huge changes to their children, sometimes even recovery from autism. I had to make sure I wasn't hallucinating, so I read that again. Could it really be true? Could diet make that much of an impact? And if this was true, why wasn't it on *20/20*? Why did moms have to find out on their own?"

McCarthy described what happened when she treated her son. "I did start vitamin B_{12} shots twice a week, and I was honestly blown away by what I saw," she wrote. "His speech doubled on the days I gave him the shots." The vitamin B_{12} shots were only the beginning. McCarthy visited a DAN doctor in San Diego who started Evan on Diflucan, a medication that kills fungi that typically live in the intestine. The results were dramatic. "Evan was still excreting yeast out of every part of his body, and every time he did, he would break through more," she wrote. "It was so liberating that I started inviting people back into my house. When my girlfriend came over, she said, 'Holy shit, Evan just had a conversation with me.' I stood in that moment and relished the confirmation of Evan's healing. It was happening before everyone's eyes. It had been about a month and a half, and Evan was still going through major yeast die-offs. According to my Google degree on yeast, this would continue until his body was a bit more balanced. I knew a big dump of yeast was about to come out of his butt when Evan looked like the Tasmanian Devil. I eventually moved everything I liked to the top shelves so it wouldn't be thrown across the room by a violent, yeast-excreting three-year-old."

. . . .

TWO DAYS LATER, ON SEPTEMBER 26, 2007, MCCARTHY APPEARED ON *Larry King Live*. This time she wasn't alone: she was joined by Jerold Kartzinel, the doctor from Melbourne, Florida, who believed MMR had caused his youngest son's autism.

King: By the way, the quick vote on our Web site. The
               question was, "Do you know someone with
               autism?" Currently, 74 percent say they do—a
               number indicating how tragically common this
               has become. Does this surprise you?

McCarthy: No, it doesn't. You know why? [Because] in
               1983 the vaccine schedule was ten: ten vac-
               cines given. Now, today, there are thirty-six,
               and a lot of people don't know that.

Later in the show, King introduced Kartzinel as "a board
certified pediatrician [whose] practice is devoted to the research
and treatment of autism and other neurodegenerative disorders."
Kartzinel had written the introduction to McCarthy's book,
where he had referred to his autistic son: "By adding cod liver
oil to his diet, we witnessed the return of eye contact and language. Autism is treatable!"

King: So what do you make of this theory we've
               been kicking around here with Jenny and the
               like with vaccines?

Kartzinel: You can't do the same thing to an entire
               population and not expect something to
               happen. For example, if you were to give
               every child in the United States a kitty cat to
               go home with, you know the majority would
               do well. But there's a small group that would
               not do well with the cats. The first thing we
               think about are allergies; they could get bit
               by the cats; the cats could run away. If we
               give every child a cat and a dog, we've got
               interactions, with the cats, the dogs, and
               between them, and you add the hedgehog
               and all of a sudden we're stacking things up
               and we can cause problems with the animals.

King: But do you not give vaccines? What do you
                 do with the vaccine if the vaccine is the
                 problem but not every child is affected by it?
                 What do you do?

Kartzinel: Well, I think we have to ask, first of all, is the
                 vaccine a problem? I keep hearing from
                 parents it is.

King: Jenny says it is.

Kartzinel: Certainly. If you tell me that your child woke
                 up with ear pain and 102 [degree] fevers and
                 I look in the ear and see an ear infection and
                 prescribe an antibiotic, you're right. If you
                 tell me that your little guy had tummy aches
                 and in the right quadrant, and he can't walk,
                 and he ends up having appendicitis, you're
                 right. Now you come in and tell me that my
                 son has lost eye contact and language and is
                 screaming all night and this happened a
                 week ago right after a vaccine, all of a
                 sudden you're wrong? I think the first thing
                 we have to understand as a medical commu-
                 nity is we have to listen to the parents tell us
                 what's going on.

McCarthy: Please listen to the parents.

McCarthy later appeared on ABC's *20/20* and *The View*. Within a week of those interviews, her book was ranked fourth on the *New York Times* nonfiction best-seller list, behind only those written by Alan Greenspan, Bill Clinton, and the family of Ron Goldman.

• • • •

AT THE TIME OF JENNY McCARTHY'S BOOK TOUR, TEN STUDIES HAD examined the relationship between MMR vaccine and autism and five between thimerosal and autism. All showed exactly the same thing: vaccines didn't cause autism. But with the help of television celebrities like Oprah Winfrey, Diane Sawyer, and Larry King, Jenny McCarthy was able to successfully counter these studies. She did it using several tried-and-true strategies.

McCarthy trumped science with personal anecdote. ("My science is Evan, and he's at home. That's my science.") Had any of these television hosts chosen to have autism experts on their show, these experts would have had to argue against a mother's personal, emotional story with statistics showing she was wrong—a nearly impossible task. The closest any of these shows came to including an expert was when Oprah Winfrey read a statement provided by the CDC. Unfortunately, the CDC—represented as a faceless, distant organization, not a caring person—didn't stand a chance.

McCarthy was persuasive not only because she was a mother who cared, but also because she was a mother who had found a cure for autism. If parents wanted to cure autism, all they had to do was remove harmful toxins and fungi from their children's bodies. ("Evan was excreting yeast out of every part of his body, and every time he did, he would break through more. When my girlfriend came over, she said, 'Holy shit, Evan just had a conversation with me.'") McCarthy's cure was reminiscent of a similar cure that had been promoted on network television several years earlier by Victoria Beck, who had claimed that secretin had cured her son. Mothers like Jenny McCarthy and Victoria Beck offer something that the physicians and scientists don't: a cure. "I think that it's very hard to defeat a wrong conviction by just saying that what [someone] believes is not true," says Harvey Fineberg, president of the Institute of Medicine.

Another strategy that clearly resonated with the television audience—and had been used with great success by Andrew Wakefield—was the conviction that doctors needed to listen to parents; therein lay truth. Jerold Kartzinel trumpeted this theme on Larry King's show by explaining that if mothers believed that vaccines caused autism, then vaccines caused autism. ("I think the first thing we have to understand as a medical community is we have to listen to the parents tell us what's going on.") Kartzinel failed to realize that doctors and scientists *have* listened to parents. That's why they had performed sixteen studies examining whether vaccines caused autism and three examining whether vaccines caused subtle developmental or psychological problems. The most recent study performed by the CDC, involving more than 1,000 children evaluated with forty-two different neurological tests, took several years to perform and cost more than $5 million. The issue for people like Jenny McCarthy isn't that doctors and scientists and public health officials haven't listened to parents; it's that they've been unable to find any evidence to validate parents' concerns.

McCarthy also appealed to the strongly held societal notion that anyone can be an expert. ("After doing my *Lorenzo's Oil* and going online and researching, I found that by fixing this, the sickness, the toxins, he was getting better.") McCarthy had trumped her pediatrician's four years of medical school, three years of residency training in pediatrics, and many years of experience practicing medicine by typing the word *autism* into Google. There she found a wealth of purported therapies her pediatrician didn't know about—therapies she believed had cured her son. She was amazed that an underground network of doctors—the only doctors who seemed to care about children with autism— was available at her fingertips. It was inconceivable to her that her pediatrician didn't know what she now knew. That the theories proposed by these doctors are varied or contradictory; that their therapies can be dangerous; that some of these doctors had been brought up before disciplinary committees for substandard medical practices; and that her doctor, far from not knowing about them, was more likely frightened by them was not something that McCarthy had considered possible.

By writing a popular book about her son's autism, Jenny McCarthy had become a media expert on vaccines. ("In 1983, the vaccine schedule was ten: ten vaccines given. Now, today, there are thirty-six, and a lot of people don't know that.") Actually, seven vaccines were routinely given to infants and young children in 1983: measles, mumps, rubella, diphtheria, tetanus, pertussis, and polio. And fourteen are given today; the additions are *Haemophilus influenzae* type b, hepatitis A, hepatitis B, rotavirus, pneumococcus, chickenpox, and influenza. But to Larry King and his producers, misstatements of fact didn't seem to matter: thirty-six vaccines, fourteen vaccines, close enough. It would have been nice if Jerold Kartzinel, described as a pediatrician and autism expert, had countered McCarthy's misinformation. But Kartzinel had been treating autistic children with alternative therapies for years. When Andrew Wakefield left England in February 2004, Kartzinel and the Good News Doctor Foundation offered him refuge in Kartzinel's International Child Development Resource Center. In addition to "curing" hundreds of children with autism, Kartzinel claims to have cured his father-in-law's Alzheimer's disease. By choosing Jenny McCarthy (an actress and model) and Jerold Kartzinel (a man who believed he had found cures for both autism and Alzheimer's disease), Larry King denied his viewers any chance to be educated about autism, its complexities, or its causes.

· · · ·

FOUR MONTHS AFTER JENNY MCCARTHY'S BOOK TOUR, THE vaccine-autism controversy again took center stage, this time with a surprising twist. On January 31, 2008, ABC aired the first episode of the television drama *Eli Stone*. Stone is a corporate lawyer who, following a hallucinatory visitation by George Michael singing "Faith," decides to represent a mother who believes her son's autism was caused by an influenza vaccine. Stone is a fictional lawyer going up against a fictional pharmaceutical company (Butel) that makes a vaccine containing a fictional preservative (mercuritol, which contains mercury). Unfortunately, the ripped-from-the-headlines story line was far from fictional.

The episode sounded several well-worn themes. First, pharmaceutical companies are evil. Stone found that Butel's CEO had insisted his daughter receive an influenza vaccine free of mercury—one made by another company (in other words, the CEO was suspicious of the mercuritol made by his own company). After the jurors heard his confession, they awarded the mother $5.2 million. Second, Stone supported the myth that some studies have found that vaccines cause autism. "Is there proof that mercuritol causes autism?" Stone asks the jury during his summation. "Yes," he says. The show never mentioned the six epidemiological studies that had clearly refuted this belief. Third, faith trumps science. During the trial, Stone visits an acupuncturist, Dr. Chen, to help relieve his hallucinations. Chen explains that Stone is going to have to make a choice. "Everything has two explanations: scientific and the divine," says Chen. "We choose which one to believe." When Chen says *divine*, he points to the setting sun, bathed in a heavenly glow. Later, Stone asks the jury to have faith even if it contradicts the evidence. "Ask yourself if you've ever believed in anything or anyone without absolute proof," he says. "That's called faith."

Although ABC's failure to accurately represent scientific studies wasn't surprising, the media's response to the episode was. When Robert F. Kennedy Jr. had claimed that federal health officials were part of a conspiracy to hide information, *Rolling Stone* magazine had been more than happy to publish the story. Similarly, when David Kirby wrote *Evidence of Harm*, or when politicians decried vaccines on Don Imus's radio program, or when parent advocacy groups took out full-page advertisements in *USA Today* and the *New York Times*, or when Jenny McCarthy promoted her book on the *Oprah Winfrey Show*, the mainstream press had always been willing to carry the story as a controversy. Not this time. Several days before the *Eli Stone* episode aired, Ed Wyatt wrote an article in the *New York Times*. Wyatt noted that ABC took "several liberties that could leave viewers believing that the debate over thimerosal—which in the script is given the fictional name mercuritol—is far from scientifically settled." But Wyatt countered that "reams of scientific studies by the leading American health authorities have failed to establish a causal link between the preservative and autism. Since the preservative was largely removed from childhood vaccines in 2001, autism rates have not declined." Choosing perspective over balance, Wyatt didn't interview David Kirby or Sallie Bernard or Lyn Redwood or Robert F. Kennedy Jr. to get the other side of a story in which scientific evidence supported only one side. He wasn't alone. Editorials in the *Boston Globe, New York Post, New York Times,* and an article in *USA Today* also questioned ABC's judgment.

Public health groups were also galvanized. Renée Jenkins, president of the AAP, wrote a letter to Anne Sweeney, president of the Disney-ABC Television Group. Jenkins knew that seventy-four children had died of influenza in 2007 and more than 300 had died in the previous four years. She mentioned this statistic in her letter and continued, "ABC will bear responsibility for the needless suffering and potential deaths of children from parents' decisions not to immunize based on the content of the [*Eli Stone*] episode." Jenkins called ABC's program "the height of reckless irresponsibility" and asked ABC to cancel the episode. If ABC insisted on showing the program, Jenkins asked the network to include a disclaimer stating that "no scientific link exists between vaccines and autism." ABC never formally responded to Jenkins by letter, e-mail, or phone, and it didn't include the disclaimer she had requested. Greg Berlanti, the cocreator and executive producer of *Eli Stone,* said he believed the episode showed both sides of the argument, and he wanted viewers to "draw their own conclusions."

Autism experts also took a stand. Nancy Minshew, director of the University of Pittsburgh's Center for Excellence in Autism, said, "The weight of evidence is so great that I don't think that there is any room for debate. I think the issue is done. I'm doing this for all the families out there who don't have a child with autism, who have to deal with the issue of 'Do I get a vaccination or do I risk my child's life' because they don't understand what the science is saying."

One lawyer also stepped forward on behalf of children. Alan Schwartz, an attorney in Columbia, Maryland, was inspired to write a letter to ABC. "Airing creative shows that stimulate discussion on controversial topics, or airing shows that are purely entertainment such as science fiction programs, is within the legitimate bounds of a free press," wrote Schwartz. "Running shows that misinform the public or titillate people to the point that they may act irresponsibly thereby causing harm—especially to children—is not. You can expect that if any child were to become seriously ill or die from a lack of inoculation in the years following airing of this episode of *Eli Stone* . . . then lawyers like myself will hold ABC responsible for the damage the television show caused."

No single episode in the vaccine-autism controversy had mobilized the public health community more than the *Eli Stone* affair.

• • • •

ON AUGUST 1, 2007, THE *AMERICAN LAWYER* OFFERED ITS READERS the lay of the land for lawsuits against the federal government and vaccine makers. "The stakes are high for autism families," wrote Elizabeth Goldberg. "Lawyers in autism cases say that the cost of treatment can run into the millions per victim." Stephen Sugarman, a professor at the School of Law at the University of California at Berkeley, saw the growing alliances against vaccines. "There are a lot of people who strongly believe in this connection, and no amount of science is going to dissuade them," he said. "They are organized. They have congressmen and celebrities on their side. And they have a group of lawyers who have now made thimerosal litigation their specialty."

On September 25, 2007, in the midst of Jenny McCarthy's book tour, Generation Rescue took out a full-page advertisement in *USA Today*. Under the headline "Are We Overvaccinating Our Kids?" the advertisement explained that children received ten vaccines in 1983 and received thirty-six vaccines today, the same numbers that Jenny McCarthy had claimed during her appearance on *Larry King Live*. The advertisement was constructed by Fenton Communications, the same group that had been hired to gain media attention for cases against silicone breast implants and Alar. J. B. Handley, Fenton Communications, Jenny McCarthy, Jerold Kartzinel, and other like-minded crusaders against vaccines were now doing their part to educate the public (which consists of potential jurors) about the harm of thimerosal. Stephen Sugarman isn't optimistic about the outcome. "Jurors' eyes gloss over when you start talking about epidemiology," said Sugarman. "Experience tells us that jurors don't trust or necessarily understand science and they are likely to make a decision completely independent of it."

The science is largely complete. Ten epidemiological studies have shown MMR vaccine doesn't cause autism; six have shown thimerosal doesn't cause autism; three have shown thimerosal doesn't cause subtle neurological problems; a growing body of evidence now points to the genes that are linked to autism; and despite the removal of thimerosal from vaccines in 2001, the number of children with autism continues to rise. Now it's up to certain parent advocacy groups, through their public relations firms, lawyers, and celebrity spokespersons, to convince the public that all of these studies are wrong—and to convince them that the doctors who proffer their vast array of alternative medicines are the only ones who really care.

NOTES

Introduction

The two best sources for information about the 1916 polio epidemic in New York City are Haven Emerson, *A Monograph on the Epidemic of Poliomyelitis (Infantile Paralysis)* (New York: Arno Press, 1977), and Naomi Rogers, *Dirt and Disease: Polio Before FDR* (New Brunswick, NJ: Rutgers University Press, 1992).

xix New York City epidemic: John Paul, *A History of Poliomyelitis* (New Haven, CT: Yale University Press, 1971); Tony Gould, *A Summer Plague: Polio and Its Survivors* (New Haven, CT: Yale University Press, 1995); Jane Smith, *Patenting the Sun: Polio and the Salk Vaccine* (New York: Morrow, 1990); Roland Berg, *Polio and Its Problems* (Philadelphia: Lippincott, 1948); Richard Carter, *Breakthrough: The Saga of Jonas Salk* (New York: Trident Press, 1966); Greer Williams, *Virus Hunters* (New York: Knopf, 1960); Aaron Klein, *Trial by Fury* (New York: Scribner's, 1972); John R. Wilson, *Margin of Safety* (New York: Doubleday, 1963); Al Burns, "The Scourge of 1916: America's First and Worst Polio Epidemic," *American Legion Magazine*, September 1966; "All United to Check Infantile Paralysis," *New York Times*, June 30, 1916; "Day Shows 12 Dead by Infant Paralysis," *New York Times*, July 2, 1916; "Bar All Children from the Movies in Paralysis War," *New York Times*, July 4, 1916; "Infantile Paralysis a Scourge and Puzzle," *New York Times*, July 9, 1916; "Believes Paralysis a Throat Infection: Dr. Bryant Says Germs Attack Only Membranes in the Air Passages of Head," *New York Times*, July 10, 1916; "31 Die of Paralysis; 162 More Ill in City," *New York Times*, July 15, 1916; "Suggests Serum for All Children: Dr. Zinghar Thinks Paralysis Might Be Checked by Using It as Preventive," *New York Times*, August 15, 1916; "Fears Subway Flies Help Spread Plague:

Edward Hatch, Jr., Asserts That Millions Breed in Refuse Left in Excavations," *New York Times*, August 18, 1916; "Drop in Paralysis Encourages City: Daily Average of New Cases in New York Falls from 173 to 131," *New York Times*, August 18, 1916; "Doctors at Odds About Paralysis; Sheffield Advises and Lovett Opposes Use of Strychnine and Massage; Adrenaline Is Recommended," *New York Times*, August 20, 1916; "Oyster Bay Revolts Over Poliomyelitis," *New York Times*, August 29, 1916; "Paralysis Defers College Openings," *New York Times*, August 31, 1916; "Meltzer Assails Auto-Inoculation; Rockefeller Institute Scientist Would Forbid Its Use in Paralysis Cases; Serum from Horse Blood; Physicians in This City See Value in Its Administration—Chickens Blamed for Epidemic," *New York Times*, September 2, 1916; "Let Children Romp as Epidemic Wanes," *New York Times*, September 5, 1916; "Believes Mosquito Spreads Paralysis: Dr. C. S. Braddock, Jr., Asserts That Careful Observations Support Theory," *New York Times*, September 9, 1916.

xix Retan: Wilson, *Margin of Safety*, 15.

1. The Tinderbox

1 Kanner study: L. Kanner, "Autistic Disturbances of Affective Contact," *Nervous Child* 2 (1943): 217–50.

1 Kanner and Donald: Ibid.

2 Ken Curtis: Cited in U.S. Congress, *Autism: Present Challenges, Future Needs*, 52.

2 Kanner and innate inability: Kanner, "Autistic Disturbances"; emphasis added.

2 Kanner and parents: Ibid.

3 Bettelheim: Grinker, *Unstrange Minds*, 78–83.

3 *The Empty Fortress*: B. Betteleheim, *The Empty Fortress: Infantile Autism and the Birth of the Self* (New York: Free Press, 1972).

4 Schreibman on behavior therapy: Schreibman, *Science and Fiction*, 133.

4 Curtis on financial burdens: Cited in U.S. Congress, *Autism: Present Challenges, Future Needs*, 52.

4 Smythe: Cited in ibid., 58.

4 Self-injurious behaviors: Schreibman, *Science and Fiction*, 47–48.

5 Matthew Israel: Cited in A. Paterson, "New Autism Treatment: Cruel or Effective?" *A Current Affair*, August 14, 2007, http://aca.ninemsn.com.au/article.aspx?id=286293.

5 Stable: Cited in A. Baker and L. Kaufman, "Autistic Boy Is Slashed to Death and His Father Is Charged," *New York Times*, November 23, 2006.

5 Lash: Cited in T. Jackman and S. McCrummen, "Ex-Bush Aide Fatally Shoots Son, Himself," *Washington Post*, July 15, 2006.

5 DeGroot: Cited in D. Reynolds, "Autistic Teen Struggled to Get Out of Blazing Apartment," http://www.raggededgemagazine.com/ide/crime_and_violence/002145.html.

5   McCarron: Cited in K. McDonald, "Mom Tried to O.D. After Killing Child," January 28, 2007, Quad-Cities Online, http://qconline.com/archives/qco/display.php?id=288998.

6   Cottrell: Cited in K. Christopher, "Autistic Boy Killed During Exorcism," http://www.findarticles.com/p/articles/mi_m2843/is_6_27/ai _110575754/print; "Autistic Boy Dies at Faith Healing Service," http://www.cnn.com/2003/US/Midwest/08/24/autistic.boy.death.

6   Biklen on facilitated communication: "Prisoners of Silence," *Frontline*, October 19, 1993.

6   Biklen and Sawyer discuss technique: "After the Storm: Free from Silence," *Primetime Live*, January 23, 1992.

7   Quotes from autistic children: Ibid.

7   Kochmeister, Smith, and Hayduke: "Prisoners of Silence," *Frontline*, October 19, 1993; "Less than a Miracle," *60 Minutes*, February 20, 1994.

8   Sawyer and awakening: "After the Storm: Free from Silence," *Primetime Live*, January 23, 1992.

8   *CBS Evening News*: "Less than a Miracle," *60 Minutes*, February 20, 1994.

8   Schreibman: Schreibman, *Science and Fiction*, 206.

8   Allen: "Facilitated Communication for Autistic Children," *All Things Considered*, National Public Radio, May 18, 1992.

9   Facilitated communication exposed: All quotes in this and the following paragraphs from Cathy Gherardi, Gerry Gherardi, Douglas Wheeler, Ray Paglieri, Marian Pitsas, Douglas Biklen, Morley Safer, and Howard Shane were obtained from "Prisoners of Silence," *Frontline*, October 19, 1993, and "Less than a Miracle," *60 Minutes*, February 20, 1994.

13   Victoria Beck and secretin: All quotes from Victoria Beck, Beck's therapist, Kenneth Sokolski, and Bernard Rimland were obtained from L. Johannes, "New Hampshire Mother Overrode Doubts on New Use of Old Drug," *Wall Street Journal*, March 10, 1999.

14   Horvath study: K. Horvath, G. Stefanatos, K. N. Sokolski et al., "Improved Social and Language Skills After Secretin Administration in Patients with Autism Spectrum Disorders," *Journal of the Association of the Academy of Minority Physicians* 9 (1998): 9–15.

15   Pauley on secretin: "Breaking the Silence," *Dateline NBC*, October 7, 1998.

15   Rimland on secretin: Cited in Fitzpatrick, *MMR and Autism*, 82.

15   Secretin price gouging and fraud: Johannes, "New Hampshire Mother Overrode Doubts."

15   Walter Herlihy: Ibid.

15   Sandler study: A. D. Sandler, K. A. Sutton, J. DeWeese et al., "Lack of Benefit of a Single Dose of Synthetic Human Secretin in the Treatment of Autism and Pervasive Developmental Disorder," *New England Journal of Medicine* 341 (1999): 1801–6.

16   Other secretin studies: P. Sturmey, "Secretin Is an Ineffective Treatment
      for Pervasive Developmental Disabilities: A Review of 15 Double-Blind
      Randomized Controlled Trials," *Research in Developmental Disabilities*
      26 (2005): 87–97.
16   Fitzpatrick on improvement: Fitzpatrick, *MMR and Autism*, 84.
17   Volkmar and secretin: F. R. Volkmar, "Lessons from Secretin," *New En-
      gland Journal of Medicine* 341 (1999): 1842–45.

2. Lighting the Fuse

The best source of information on the MMR-autism controversy is Fitzpatrick,
MMR and Autism. Unless otherwise indicated, all quotes from Dan Burton,
Dennis Kucinich, Shelly Reynolds, Jeanna Smith, Scott Bono, Andrew Wake-
field, John O'Leary, Mary Megson, Michael Goldberg, Vijendra Singh, John
Upledger, Brent Taylor, and Henry Waxman were obtained from U.S. Congress,
Autism: Present Challenges, Future Needs.

19   Horton on Wakefield: Horton, *MMR Science*, 22
19   Wakefield's Crohn's, measles paper: A. J. Wakefield, R. M. Pittilo, R. Sim
      et al., "Evidence of Persistent Measles Virus Infection in Crohn's Disease,"
      Journal of Medical Virology 39 (1993): 345–53.
19   Wakefield's Crohn's, measles vaccine paper: N. P. Thompson, S. M. Mont-
      gomery, R. E. Pounder, and A. J. Wakefield, "Is Measles Vaccination a Risk
      Factor for Inflammatory Bowel Disease?" *Lancet* 345 (1995): 1071–74.
19   Failure to confirm Wakefield's Crohn's studies: P. Farrington and E. Miller,
      "Measles Vaccination as a Risk Factor for Inflammatory Bowel Disease,"
      Lancet 345 (1995): 1362; T. Gilat, D. Hacohen, P. Lilos, and M. J. S.
      Langman, "Childhood Factors in Ulcerative Colitis and Crohn's Disease:
      An International Co-Operative Study," *Scandinavian Journal of Gastro-
      enterology* 22 (1987): 1009–24; J. Hermon-Taylor, J. Ford, N. Sumar et
      al., "Measles Virus and Crohn's Disease," *Lancet* 345 (1995): 922–23;
      M. Iizuka, O. Nakagomi, M. Chiba et al., "Absence of Measles Virus in
      Crohn's Disease," *Lancet* 345 (1995): 199; M. Feeney, A. Clegg, P. Win-
      wood, and J. Snook, "A Case-Control Study of Measles Vaccination and
      Inflammatory Bowel Disease," *Lancet* 350 (1997): 764–66; P. Daszak, M.
      Purcell, J. Lewin et al., "Detection and Comparative Analysis of Persistent
      Measles Virus Infection in Crohn's Disease by Immunogold Electron Mi-
      croscopy," *Journal of Clinical Pathology* 50 (1997): 299–304; N. C. Fisher,
      L. Yee, P. Nightingale et al., "Measles Virus Serology in Crohn's Disease,"
      Gut 41 (1997): 66–69; D. S. Pardi, W. J. Tremaine, W. J. Sandborn et al.,
      "Perinatal Exposure to Measles Virus Is Not Associated with the Devel-
      opment of Inflammatory Bowel Disease," *Inflammatory Bowel Diseases* 5
      (1999): 104–6; M. Iizuka, M. Chiba, M. Yukawa et al., "Immunohis-
      tochemical Analysis of the Distribution of Measles Related Antigen in the
      Intestinal Mucosa in Inflammatory Bowel Disease," *Gut* 46 (2000): 163–69;

D. L. Morris, S. M. Montgomery, N. P. Thompson et al., "Measles Vaccination and Inflammatory Bowel Disease: A National British Cohort Study," *American Journal of Gastroenterology* 95 (2000): 3507–12; M. Iizuka, H. Saito, M. Yukawa et al., "No Evidence of Persistent Mumps Virus Infection in Inflammatory Bowel Disease," *Gut* 48 (2001): 637–41.

19   Wakefield retracts Crohn's findings: N. Chadwick, I. J. Bruce, S. Schepelmann et al., "Measles Virus RNA Is Not Detected in Inflammatory Bowel Disease Using Hybrid Capture and Reverse Transcription Followed by the Polymerase Chain Reaction," *Journal of Medical Virology* 55 (1998): 305–11.

19   Wakefield on hypothesis testing: A. J. Wakefield, "MMR Vaccination and Autism," *Lancet* 354 (1999): 949.

20   Wakefield *Lancet* paper: A. J. Wakefield, S. H. Murch, A. Anthony et al., "Ileal-Lymphoid-Nodular Hyperplasia, Non-Specific Colitis, and Pervasive Developmental Disorder in Children," *Lancet* 351 (1998): 637–41.

20   Wakefield and listening to parents: A. J. Wakefield, "The Case Against MMR: Wary Parents Have Proved the Experts Wrong Before. They Will Do So Again," *The Independent* (London), January 22, 2001.

20   Isabella Thomas: Cited in Horton, *MMR Science*, 9.

21   Wakefield on separating MMR: Cited in B. Deer, "MMR: The Truth Behind the Crisis," *Sunday Times* (London), February 22, 2004.

22   Murch: Cited in Fitzpatrick, *MMR and Autism*, 18.

22   Zuckerman: Cited in S. Boseley, "Scientists Go Public with Doubts Over MMR Vaccine," *The Guardian*, February 27, 1998.

22   *Hear the Silence*: M. Wells, "Five Plans Autism Drama," *The Guardian*, May 23, 2003; M. Wells and S. Boseley, "Calls to Axe TV Drama on MMR," *The Guardian*, December 3, 2003; S. Boseley, "Doctors Boycott Debate to Follow MMR Drama," *The Guardian*, December 4, 2003; D. Aaronovitch, "Comment: A Travesty of Truth," *The Observer*, December 14, 2003; J. Stevenson, "I'm a Mother, Too," *The Independent* (London), December 14, 2003; M. Taylor, "Mothers Alarmed After TV MMR Drama," *The Guardian*, December 16, 2003.

23   *Hear the Silence* quotes: All quotes from the television docudrama were obtained from Aaronovitch, "Comment."

23   Bruce: Cited in M. Taylor, "Mothers Alarmed After TV MMR Drama," *The Guardian*, December 16, 2003.

23   Mitchell: Cited in Fitzpatrick, *MMR and Autism*, 58.

24   Measles outbreaks: N. Gould, "The Town Divided by a Deadly Disease," *Belfast Telegraph*, November 14, 2004; "Fall in MMR Vaccine Coverage Reported as Further Evidence of Vaccine Safety Is Published," *CDR Weekly*, June 25, 1999; B. Lavery, "As Vaccination Rates Decline in Ireland, Cases of Measles Soar," *New York Times*, February 8, 2003; T. Peterkin, "Alert Over 60 Percent Rise in Measles," *London Daily Telegraph*, May 12, 2003; N. Begg, M. Ramsey, J. White, and Z. Bozoky, "Media

Dents Confidence in MMR Vaccine," *British Medical Journal* 316 (1998): 561; B. Deer, "Schoolboy, 13, Dies as Measles Makes a Comeback," *Sunday Times* (London), April 2, 2006; R. Boyles and A. Browne, "MMR Jab Urged After 420 Pupils Are Struck by Measles," *The Times* (London), April 6, 2006; K. Mansey, "MMR Link to Mumps Cases," *Daily Post*, January 16, 2006; S. Boseley, "MMR Vaccinations Fall to New Low," *The Guardian*, September 24, 2004; E. K. Mulholland, "Measles in the United States, 2006," *New England Journal of Medicine* 355 (2006): 440–43; J. McBrien, J. Murphy, D. Gill et al., "Measles Outbreak in Dublin," *Pediatric Infectious Disease Journal* 22 (2003): 580–84; P. A. Brunell, "More on Measles and the Impact of the *Lancet* Retraction," *Infectious Diseases in Children*, May 2004; B. Deer, "MMR Scare Doctor Faces List of Charges," *The Times* (London), September 11, 2005; S. Hastings, "Doctor at Sharp End of MMR Controversy," *Yorkshire Post*, June 14, 2006.

24　Naomi's death: *Fragile Immunity*, video produced by PATH, narrated by Ian Holm, 2004.

24　*Science* paper: V. A. A. Jansen, N. Stollenwerk, H. J. Jensen et al., "Measles Outbreaks in a Population with Declining Vaccine Uptake," *Science* 301 (2003): 804.

24　Measles death in England: B. Deer, "Schoolboy, 13, Dies."

25　O'Leary: Cited in V. Uhlmann, C. M. Martin, O. Sheils et al., "Potential Viral Pathogenic Mechanism for New Variant Inflammatory Bowel Disease," *Journal of Clinical Pathology* 55 (2002): 84–90.

25　Kawashima: H. Kawashima, T. Mori, Y. Kasiwagi et al., "Detection and Sequencing of Measles Virus from Peripheral Mononuclear Cells from Patients with Inflammatory Bowel Disease and Autism," *Digestive Diseases and Sciences* 45 (2000): 723–29.

25　Singh: V. K. Singh, S. X. Lin, E. Newell, and C. Nelson, "Abnormal Measles-Mumps-Rubella Autoantibodies and CNS Autoimmunity in Children with Autism," *Journal of Biomedical Science* 9 (2002): 359–64; V. K. Singh and R. L. Jensen, "Elevated Levels of Measles Antibodies in Children with Autism," *Pediatric Neurology* 28 (2003): 292–94; V. K. Singh, S. X. Lin, and V. C. Yang, "Serological Association of Measles Virus and Human Herpesvirus-6 with Brain Autoantibodies in Autism," *Clinical Immunology and Immunopathology* 89 (1998): 105–9.

25　Aitken: H. Mills, "MMR: The Story So Far," *Private Eye*, May 2002.

25　Spitzer: W. O. Spitzer, K. J. Aitken, S. Dell'Anniello, and M. W. L. Davis, "The Natural History of Autistic Syndrome in British Children Exposed to MMR," *Adverse Drug Reactions and Toxicological Reviews* 20 (2001): 47–55.

26　March: Mills, "MMR."

26　Kinsbourne and Menkes: Fitzpatrick, *MMR and Autism*, 116.

26　Krigsman: Cited in L. Fraser, "U.S. Experts Back MMR Doctor's Findings," *Sunday Telegraph*, June 23, 2002.

26 Burton and Laetrile: B. Wilson, "The Rise and Fall of Laetrile," http://
    www.quackwatch.org/01QuackeryRelatedTopics/Cancer/laetrile.html.
27 Burton and Ephedra: Cited in S. Brownlee, "Swallowing Ephedra," http://
    archive.salon.com/health/feature/2000/06/07/ephedra.
27 Burton and AIDS: E. Walsh, "Burton: A 'Pit Bull' in the Chair," *Washing-
    ton Post*, March 19, 1997; F. Pellegrini, "Fool on the Hill," *Time.com*,
    http://www.time.com/time/daily/special/look/burton.
36 American children and MMR vaccine: M. J. Smith, L. M. Bell, S. E.
    Ellenberg, and D. M. Rubin, "Media Coverage of the MMR-Autism Con-
    troversy and Its Relationship to MMR Immunization Rates in the United
    States," *Pediatrics* 121 (2008): e836–e843.
36 Fitzpatrick and butterfly effect: Fitzpatrick, *MMR and Autism*, x.

3. The Implosion

The best sources for the fall of Andrew Wakefield and his theory are B. Deer,
"MMR: The Truth Behind the Crisis," *Sunday Times* (London), February 22,
2004, and Fitzpatrick, *MMR and Autism*.

37 Horton and Wakefield meeting: Horton, *MMR Science*, 2.
37 Ethical Practices Committee: A. J. Wakefield, S. H. Murch, A. Anthony
    et al., "Ileal-Lymphoid-Nodular Hyperplasia, Non-Specific Colitis, and Per-
    vasive Developmental Disorder in Children," *Lancet* 351 (1998): 637–41.
37 Colon perforation: R. Ellis, "£500,000 Payout for Autistic Boy Left Fight-
    ing for Life After Being Used as an MMR Guinea Pig," *Daily Mail*, De-
    cember 8, 2007.
38 Wakefield study funding: Wakefield, Murch, Anthony et al., "Ileal-
    Lymphoid-Nodular Hyperplasia."
38 Horton on Wakefield bias: Horton, *MMR Science*, 3–4.
38 Kessick and illness: Cited in H. Mills, "MMR: The Story So Far," *Private
    Eye*, May 2002.
38 Kessick and worry: Cited in G. Frankel, "Charismatic Doctor at Vortex of
    Vaccine Dispute," *Washington Post*, July 11, 2004.
39 Kessick and broken hearts: Cited in G. Langdon-Down, "Law: A Shot in
    the Dark," *The Independent*, November 27, 1996.
39 Kessick and Wakefield: Cited in Frankel, "Charismatic Doctor."
39 Kessick visits JABS: Langdon-Down, "Law."
39 Barr and Opren: J. Erlichman, "Lawyers of Opren Victims in Payout,"
    The Guardian, December 1, 1987; R. Hughes and P. Marsh, "Lawyers in
    Opren Cases May Be in Agreement," *Financial Times*, December 9, 1987;
    J. Erlichman, "Average of 2,000 Pounds Offered to Opren Victims," *The
    Guardian*, December 10, 1987; C. Dyer, "Opren Firm Offer Aims to Stifle
    Expert Solicitors," *The Guardian*, December 14, 1987; R. Hughes, "Late
    Opren Claimants Face Compensation Fight," *Financial Times*, December

17, 1987; C. Dyer, "Opren Claimants' Hopes Dashed," *The Guardian*, February 1, 1991.

39   Barr and worst day: Cited in L. Tsang, "If You Decide to Become a Lawyer, Do Something Else First to Get Life Skills," *The Times* (London), February 26, 2002.

39   Barr and organophosphates: M. Fitzpatrick, "Medicine on Trial," *Spiked*, December 15, 2003.

40   Barr and other law firms: Fitzpatrick, *MMR and Autism*, 104.

40   Barr and MMR cases: T. Thompson, "Parents Seek Checks Over Vaccine Fears," *The Scotsman*, April 26, 1999; Mills, "MMR."

40   Barr and MMR theory: Cited in Fitzpatrick, *MMR and Autism*, 101.

40   Horton and authors: Horton, *MMR Science*, 5.

40   Horton on BBC: Ibid., 6.

40   Horton and "fatal conflicts": Ibid.

41   Wakefield confronted by reporters: Cited in Deer, "MMR."

41   Wakefield response to funding controversy: Horton, *MMR Science*, 5.

41   Barr on funding controversy: Cited in Deer, "MMR."

41   Murch and "unpleasant surprise": Ibid.

41   *Lancet* retraction: Cited in Horton, *MMR Science*, 12; emphasis added.

42   O'Leary shocked: Cited in Frankel, "Charismatic Doctor."

42   Wakefield caveat: Wakefield, Murch, Anthony et al., "Ileal-Lymphoid-Nodular Hyperplasia."

42   Studies exonerating MMR: B. Taylor, E. Miller, C. P. Farrington et al., "Autism and Measles, Mumps, and Rubella Vaccine: No Epidemiological Evidence for a Causal Association," *Lancet* 353 (1999): 2026–29; H. Peltola, A. Patja, P. Leinikki et al., "No Evidence for Measles, Mumps, and Rubella Vaccine-Associated Inflammatory Bowel Disease or Autism in a 14-Year Prospective Study," *Lancet* 351 (1998): 1327–28; F. DeStefano and R. T. Chen, "Negative Association Between MMR and Autism," *Lancet* 353 (1999): 1986–87; E. Fombonne, "Are Measles Infections or Measles Immunizations Linked to Autism?" *Journal of Autism and Developmental Disorders* 29 (1999): 349–50; J. A. Kaye, M. Melero-Montes, and H. Jick, "Mumps, Measles, and Rubella Vaccine and the Incidence of Autism Recorded by General Practitioners: A Time Trend Analysis," *British Medical Journal* 322 (2001): 460–63; R. L. Davis, P. Kramarz, B. Kari et al., "Measles-Mumps-Rubella and Other Measles-Containing Vaccines Do Not Increase the Risk for Inflammatory Bowel Disease: A Case-Control Study from the Vaccine Safety DataLink Project," *Archives of Pediatrics and Adolescent Medicine* 155 (2002): 354–59; L. Dales, S. J. Hammer, and N. J. Smith, "Time Trends in Autism and in MMR Immunization Coverage in California," *Journal of the American Medical Association* 285 (2001): 1183–85; C. P. Farrington, E. Miller, and B. Taylor, "MMR and Autism: Further Evidence Against a Causal Association," *Vaccine* 19 (2001): 3632–35; E. Fombonne and S. Chakrabarti, "No Evidence for a New Vari-

ant of Measles-Mumps-Rubella-Induced Autism," *Pediatrics* 108 (2001), http://www.pediatrics.org/cgi/content/full.108/4/e58; N. A. Halsey and S. L. Hyman, "Measles-Mumps-Rubella Vaccine and Autistic Spectrum Disorder: Report from the New Challenges in Childhood Immunization Conference Convened in Oak Brook, Illinois, June 12, 2000," *Pediatrics* 107 (2001), http://www.pediatrics.org/cgi/content/full/107/5/e84; B. Taylor, E. Miller, R. Lingam et al., "Measles, Mumps, and Rubella Vaccination and Bowel Problems or Developmental Regression in Children with Autism: Population Study," *British Medical Journal* 324 (2002): 393–96; K. M. Madsen, A. Hviid, M. Vestergaard et al., "A Population-Based Study of Measles, Mumps, and Rubella Vaccination and Autism," *New England Journal of Medicine* 347 (2002): 1477–82; A. Mäkela, J. P. Nuorti, and H. Peltola, "Neurologic Disorders After Measles-Mumps-Rubella Vaccination," *Pediatrics* 110 (2002): 957–63; P. A. Offit and S. E. Coffin, "Communicating Science to the Public: MMR Vaccine and Autism," *Vaccine* 22 (2003): 1–6; F. DeStefano, T. K. Bhasin, W. W. Thompson et al., "Age at First Measles-Mumps-Rubella Vaccination in Children with Autism and School-Matched Control Subjects: A Population-Based Study in Metropolitan Atlanta," *Pediatrics* 113 (2004): 259–66; E. Fombonne and E. H. Cook Jr., "MMR and Autistic Enterocolitis: Consistent Epidemiological Failure to Find an Association," *Molecular Psychiatry* 8 (2003): 133–34; K. Wilson, E. Mills, C. Ross, J. McGowan et al., "Association of Autistic Spectrum Disorder and the Measles, Mumps, and Rubella Vaccine," *Archives of Pediatric and Adolescent Medicine* 157 (2003): 628–34; H. Honda, Y. Shimizu, and M. Rutter, "No Effect of MMR Withdrawal on the Incidence of Autism: A Total Population Study," *Journal of Child Psychiatry and Psychology* 46 (2005): 572–79.

43   Researchers fail to find chronic measles infection: Y. D'Souza, E. Fombonne, and B. J. Ward, "No Evidence of Persisting Measles Virus in Peripheral Blood Mononuclear Cells from Children with Autism Spectrum Disorder," *Pediatrics* 118 (2006): 1664–75; M. A. Afzal, L. C. Ozoemena, A. O'Hare et al., "Absence of Detectable Measles Virus Genome Sequence in Blood of Autistic Children Who Have Had Their MMR Vaccination During the Routine Childhood Immunization Schedule of UK," *Journal of Medical Virology* 78 (2006): 623–30; G. Baird, A. Pickles, E. Simonoff et al., "Measles Vaccination and Antibody Response in Autism Spectrum Disorders," *Archives of Diseases of Children* (2008) [in press].

43   Ward: Cited in M. Rauscher, "New Data Refute Measles Virus Persistence in Children with Autism," *Reuters*, October 11, 2006.

43   Poor performance of Unigenetics: Cited in S. L. Katz, "Has the Measles-Mumps-Rubella Vaccine Been Fully Exonerated?" *Pediatrics* 118 (2006): 1744–45.

44   Plotkin: Stanley Plotkin, personal communication, 2004.

44   O'Leary: Cited in Fitzpatrick, *MMR and Autism*, 127.

44   MMR and bowel disease: DeStefano and Chen, "Negative Associa-
     tion Between MMR and Autism"; Davis, Kramarz, Kari et al.,
     "Measles-Mumps-Rubella and Inflammatory Bowel Disease"; Tay-
     lor, Miller, Lingam et al., "Measles, Mumps, and Rubella in Children
     with Autism."

44   Bowel problems: R. T. Chen and F. DeStefano. "Vaccine Adverse Events:
     Causal or Coincidental," *Lancet* 351 (1998): 611–12.

44   Gershon and leak: Cited in Fitzpatrick, *MMR and Autism*, 146.

44   Vaccine challenge: P. A. Offit, J. Quarles, M. A. Gerber et al., "Addressing
     Parents' Concerns: Do Multiple Vaccines Overwhelm or Weaken the In-
     fant's Immune System?" *Pediatrics* 109 (2002): 124–29.

45   Singh challenged: Ben Schwartz and Bill Bellini, CDC, personal communi-
     cation.

45   Horton and affair "coming to a head": Horton, *MMR Science*, 2–3.

46   Deer and Freedom of Information Act: "Freedom of Information Act Re-
     quest: MMR/MR Multi-Party Action," Secretariat, Legal Services Com-
     mission, December 22, 2006.

47   Medical investigation team: Fitzpatrick, *MMR and Autism*, 114; Deer,
     "MMR."

47   Wakefield and nondenial: Cited in Horton, *MMR Science*, 5; A. Wake-
     field. "MMR—Responding to Retraction," *Lancet* 363 (2004): 1327–28.

47   Wakefield patent: "Pharmaceutical Composition for Treatment of IBD
     and RBD," patent application number 9711663.6, filed June 6, 1997.

48   MacDonald: B. Deer, "MMR Scare Doctor Planned Rival Vaccine," *Sun-
     day Times* (London), November 14, 2004.

48   O'Leary and Unigenetics: Cited in B. Deer, http://briandeer.com/mmr/
     oleary-statement.htm.

48   Aitken book: K. Aitken et al., *Children with Autism: Diagnosis and Inter-
     ventions to Meet Their Needs* (London: Jessica Kingsley, 1998).

49   Aitken resigns: B. Christie, "Child Psychologist Quits in Porn Shame,"
     Sunday Times (London), November 29, 1998; "Porn Psychologist Quits,"
     The Herald (Glasgow), November 30, 1998; D. Clarke, "Top Children's
     Doctor Quits Hospital After Porn Found in His Office," *Daily Mail*, No-
     vember 30, 1998.

49   Spitzer and MacDonald: Cited in http://briandeer.com/mmr.

50   March and Damascene conversion: M. Fitzpatrick, personal communica-
     tion, January 5, 2008.

50   Fitzpatrick on March money: Ibid.

50   March and funding: Cited in B. Deer, "MMR Doctor Given Legal Aid
     Thousands," *Sunday Times* (London), December 31, 2006.

50   Harris: Ibid.

50   Barr and defense of team: R. Barr, "The Withdrawal of Legal Aid in the
     MMR Cases," *Solicitors Journal*, October 2003.

50   Autistic children in Detroit: Ibid.

51   Legal Services Commission withdrawal: N. Martin, "Parents Seeking MMR Compensation Lose Legal Aid for Court Fight," *Daily Telegraph*, October 2, 2003.

51   Dodgson: Cited in ibid.

51   Barr and Venus de Milo: Barr, "The Withdrawal of Legal Aid."

52   Medical truths: Cited in Fitzpatrick, "Medicine on Trial."

52   Wakefield steps down: Cited in L. Fraser, "Anti-MMR Doctor Is Forced Out," *Sunday Telegraph*, December 2, 2001.

52   Wakefield in Florida: A. Ahuja, "MMR Maverick," *The Times* (London), June 13, 2006; Fitzpatrick, *MMR and Autism*, 163; Horton, *MMR Science*, 131.

52   Wakefield and GMC: Deer, http://briandeer.com/mmr; S. Boseley, "Doctor Behind MMR Scare to Face Four Charges of Misconduct Over Research," *The Guardian*, June 11, 2006.

53   Wakefield and *60 Minutes*: "The MMR Vaccine," *60 Minutes*, November 12, 2000.

54   Wakefield on quitting: Cited in Deer, "MMR."

54   Fitzpatrick on Wakefield: Fitzpatrick, *MMR and Autism*, 160.

55   Fitzpatrick and MMR anxieties: M. Fitzpatrick, personal communication, January 5, 2008.

55   Wakefield and proof: Wakefield, Murch, Anthony et al., "Ileal-Lymphoid-Nodular Hyperplasia."

55   Salisbury on Wakefield: D. Salisbury, interview, January 9, 2008.

56   Fitzpatrick on Royal Free Hospital: M. Fitzpatrick, personal communication, January 5, 2008.

56   Boyce and media: Cited in M. Fitzpatrick, "Anti-MMR Mania: Diagnoses and Cure," http://www.spiked-online.com/index.php?/site/reviewofbooks _article/4362; Boyce, *Health, Risk and News*.

56   Boyce and balance: Ibid.

57   *Lancet*: S. Connor, "It Is Britain's Pre-Eminent Medical Journal: Now Its Reputation Hangs on a Single Issue," *The Independent*, October 15, 1999.

57   Martin on Horton: N. Martin. "A Pariah and a Firebrand in the Eye of the Storm," *Daily Telegraph*, February 23, 2004.

57   Journalist on Horton: Cited in ibid.

57   Chen and DeStefano: Chen and DeStefano, "Vaccine Adverse Events."

57   Horton and censorship: Horton, *Second Opinion*, 320.

58   Horton and arrogance: Ibid., 214.

58   Salisbury on Horton: D. Salisbury, interview, January 9, 2008.

58   Horton and genetically modified foods: S. W. Ewen and A. Pusztai, "Effect of Diets Containing Genetically Modified Potatoes Expressing Galanthus Nivalis Lectin on Rat Small Intestine," *Lancet* 354 (1999): 1353–54; S. W. Ewen and A. Pusztai, "Health Risks of Genetically Modified Foods," *Lancet* 354 (1999): 684; Connor, "It Is Britain's Pre-Eminent Medical Journal."

58 Horton and silicone breast implants: S. A. Tenenbaum, J. C. Rice, L. R. Espinosa et al., "Use of Antipolymer Antibody Assay in Recipients of Silicone Breast Implants," *Lancet* 349 (1997): 449–54.

59 *Lancet* and Iraq war casualties: G. Burnham, R. Lafta, S. Doocy, and L. Roberts, "Mortality After the 2003 Invasion of Iraq: A Cross-Sectional Cluster Survey," *Lancet* 386 (2006): 1421–28; "The Lancet's Political Hit," *Wall Street Journal*, January 9, 2008; C. A. Brownstein and J. S. Brownstein, "Estimating Excess Mortality in Post-Invasion Iraq," *New England Journal of Medicine* 358 (2008): 445–47.

4. A Precautionary Tale

61 FDA Modernization Act: "Uproar Over a Little-Known Preservative, Thimerosal, Jostles U.S. Hepatitis B Vaccination Policy," *Hepatitis Control Report* 4 (1999): 3–7; Allen, *Vaccine*, 377.

61 FDA and mercury compounds: L. K. Ball, R. Ball, and R. D. Pratt, "An Assessment of Thimerosal Use in Childhood Vaccines," *Pediatrics* 107 (2001): 1147–54; S. A. Plotkin, "Report on Meeting Called by the American Academy of Pediatrics to Discuss Thimerosal in Vaccines," June 30, 1999, personal communication; G. L. Freed, M. C. Andreae, A. E. Cowan, and S. L. Katz, "The Process of Public Policy Formulation: The Case of Thimerosal in Vaccines," *Pediatrics* 109 (2002): 1153–59.

61 Minamata Bay: P. W. Davidson, G. J. Myers, and B. Weiss, "Mercury Exposure and Child Development Outcomes," *Pediatrics* 113 (2004): 1023–29; M. Harada, "Minamata Disease: Methylmercury Poisoning in Japan Caused by Environmental Pollution," *Critical Reviews in Toxicology* 25 (1995): 1–24.

62 Contamination of vaccines: Ball, Ball, and Pratt, "An Assessment of Thimerosal."

63 Lilly: Ibid.

63 Mercury safety guidelines: Institute of Medicine, *Immunization Safety Review: Thimerosal-Containing Vaccines and Neurodevelopmental Disorders.*

64 Halsey and disbelief: Cited in A. Allen, "The Not-So-Crackpot Autism Theory," *New York Times Magazine*, November 10, 2002.

65 Halsey in Maine: Ibid.

65 Reassuring studies: D. O. Marsh, T. W. Clarkson, C. Cox et al., "Fetal Methylmercury Poisoning: Relationship Between Concentration in Single Strands of Maternal Hair and Child Effects," *Archives of Neurology* 44 (1987): 1017–22; L. Magos, A. W. Brown, S. Sparrow et al., "The Comparative Toxicology of Ethyl- and Methyl-Mercury," *Archives of Toxicology* 57 (1985): 260–67; L. Magos, "Review on the Toxicity of Ethylmercury, Including Its Presence as a Preservative in Biological and Pharmaceutical Products," *Journal of Applied Toxicology* 21 (2001): 1–5; P. W. Davidson,

G. J. Myers, C. Cox et al., "Effects of Prenatal and Postnatal Methylmercury Exposure from Fish Consumption on Neurodevelopment: Outcomes at 66 Months of Age in the Seychelles Child Development Study," *Journal of the American Medical Association* 280 (1998): 701–7; P. W. Davidson, D. Palumbo, G. J. Myers et al., "Neurodevelopmental Outcomes of Seychellois Children from the Pilot Cohort at 108 Months Following Prenatal Exposure to Methylmercury from a Maternal Fish Diet," *Environmental Research Section A* 84 (2000): 1–11; M. E. Pichichero, A. Gentile, N. Giglio et al., "Mercury Levels in Newborns and Infants After Receipt of Thimerosal-Containing Vaccines," *Pediatrics* 121 (2008): e208–e214.

66   Faroe Islands study: P. Grandjean, P. Weihe, R. F. White et al., "Cognitive Deficit in 7-Year-Old Children with Prenatal Exposure to Methylmercury," *Neurotoxicology and Teratology* 19 (1997): 417–28.

67   Halsey teleconferences: Interviews with those who participated in the teleconferences in June 1999 were obtained from Jon Abramson on October 19, 2007; Carol Baker on November 12, 2007; Meg Fisher on October 18, 2007; John Modlin on October 16, 2007; Martin Myers on November 14, 2007; Walter Orenstein on October 17, 2007; Georges Peter on October 18, 2007; and Larry Pickering on October 31, 2007.

71   AAP, PHS joint statement: Public Health Service and the American Academy of Pediatrics, "Thimerosal in Vaccines: A Joint Statement of the American Academy of Pediatrics and the Public Health Service," *Morbidity and Mortality Weekly Report* 48 (1999): 563–65.

71   AAP press release: American Academy of Pediatrics, July 7, 1999.

72   Music: S. Music, "Sad Reflections on a Thimerosal Workshop at the NIH," August 12, 1999, personal communication.

72   Plotkin: Cited in Allen, *Vaccine*, 382.

72   CDC warning: Centers for Disease Control and Prevention, "Thimerosal and Vaccines: Talking Points," July 5, 1999.

73   October meeting: Minutes from Advisory Committee on Immunization Practices meeting, October 20, 1999.

73   Breast implant story: Unless otherwise stated, details of the silicone breast implant controversy were obtained from Angell, *Science on Trial*.

74   Breast implants and connective tissue diseases: S. A. Van Nunen, "Post-Mammoplasty Connective Tissue Disease," *Arthritis and Rheumatism* 25 (1982): 694–97.

74   Kessler: Cited in Angel, *Science on Trial*, 50.

75   Gabriel: S. E. Gabriel, M. O'Fallon, L. T. Kurland et al., "Risk of Connective-Tissue Diseases and Other Disorders After Breast Implantation," *New England Journal of Medicine* 330 (1994): 1697–702.

76   Sánchez-Guerrero: J. Sánchez-Guerrero, G. A. Colditz, E. W. Karlson et al., "Silicone Breast Implants and the Risk of Connective Tissue Diseases and Symptoms," *New England Journal of Medicine* 332 (1995): 1666–70.

76 Studies exonerating breast implants: C. J. Burns, T. J. Laing, B. W. Gillespie et al., "The Epidemiology of Scleroderma Among Women: Assessment of Risk from Exposure to Silicone and Silica," *Journal of Rheumatology* 23 (1996): 1904–11; S. M. Edworthy, L. Martin, S. G. Barr et al., "A Clinical Study of the Relationship Between Silicone Breast Implants and Connective Tissue Disease," *Journal of Rheumatology* 25 (1998): 254–60; F. Wolfe and J. Anderson, "Silicone Filled Breast Implants and the Risk of Fibromyalgia and Rheumatoid Arthritis," *Journal of Rheumatology* 26 (1999): 2025–28; J. A. Goldman, J. Greenblatt, R. Joines et al., "Breast Implants, Rheumatoid Arthritis, and Connective Tissue Diseases in a Clinical Practice," *Journal of Clinical Epidemiology* 48 (1995): 571–82; B. L. Strom, M. M. Reidenberg, B. Freundlich, and R. Schinnar, "Breast Silicone Implants and Risk of Systemic Lupus Erythematosus," *Journal of Clinical Epidemiology* 47 (1994): 1211–14; H. Hennekens, I. Lee, N. R. Cook et al., "Self-Reported Breast Implants and Connective Tissue Diseases in Female Health Professionals," *Journal of the American Medical Association* 275 (1996): 616–21.

77 Kossovsky study: N. Kossovsky, J. P. Heggers, and M. C. Robson, "Experimental Demonstration of the Immunogenicity of Silicone-Protein Complexes," *Journal of Biomedical Materials Research* 21 (1987): 1125–33.

77 Studies on silicone antibodies and breast implants: S. H. Lamm, "Antipolymer Antibodies, Silicone Breast Implants, and Fibromyalgia," *Lancet* 349 (1997): 1170–71; S. A. Edlavitch, "Antipolymer Antibodies, Silicone Breast Implants, and Fibromyalgia," *Lancet* 349 (1997): 1170; J. H. Korn. "Antipolymer Antibodies, Silicone Breast Implants, and Fibromyalgia," *Lancet* 349 (1997): 1171; M. Angell, "Antipolymer Antibodies, Silicone Breast Implants, and Fibromyalgia," *Lancet* 349 (1997): 1172; T. M. Ellis, N. S. Hardt, and M. A. Atkinson, "Antipolymer Antibodies, Silicone Breast Implants, and Fibromyalgia," *Lancet* 349 (1997): 1173; S. H. Miller, "Silicone Breast Implants and Antipolymer Antibodies," *Lancet* 350 (1997): 740.

78 IOM report: C. Marwick, "Are They Real? IOM Report on Breast Implant Problems," *Journal of the American Medical Association* 282 (1999): 314–15.

78 Gabriel: Cited in G. Kolata, "Legal System and Science Coming to Different Conclusions on Silicone," *New York Times*, May 16, 1995.

78 Angell: Cited in ibid.

79 *Wall Street Journal* editorial: "Implants and Science," *Wall Street Journal*, November 20, 2006.

79 Harm from delaying hepatitis B vaccine: M. E. Tucker, "New Hepatitis B Vaccine Guidelines Cause Confusion," *Pediatric News* 33 (1999): 1–4; M. B. Hurie, T. N. Saari, and J. P. Davis, "Impact of the Joint Statement by the American Academy of Pediatrics/U.S. Public Health Service on Thimerosal in Vaccines on Hospital Infant Hepatitis B Vaccination Practices," *Pediatrics* 107 (2001): 755–58; "Uproar Over a Little-Known Preserva-

tive, Thimerosal, Jostles U.S. Hepatitis B Vaccination Policy," *Hepatitis Control Report* 4 (1999): 3–7; R. M. Brayden, K. A. Pearson, J. S. Jones et al., "Effect of Thimerosal Recommendations on Hospitals' Neonatal Hepatitis B Vaccination Policies," *Journal of Pediatrics* 138 (2001): 752–55; R. Thomas, A. E. Fiore, H. L. Corwith et al., "Hepatitis B Vaccine Coverage Among Infants Born to Women Without Prenatal Screening for Hepatitis B Virus Infection: Effects of the Joint Statement on Thimerosal in Vaccines," *Pediatric Infectious Diseases Journal* 23 (2004): 313–18; "Impact of the 1999 AAP/USPHS Joint Statement on Thimerosal in Vaccines on Infant Hepatitis B Vaccination Practices," *Morbidity and Mortality Weekly Report* 50 (2001): 94–97.

80　Halsey to thank: Allen, "The Not-So-Crackpot Autism Theory."

5. Mercury Rising
The epigraph in this chapter is from "The Rise Against Mercury," *Seed Magazine*, May 2004.

81　Redwood and son: Cited in U.S. Congress, House of Representatives, Committee on Government Reform, *Mercury in Medicine: Are We Taking Unnecessary Risks?* 106th Congress, Second session, July 18, 2000 (Washington, DC: U.S. Government Printing Office, 2001), 12.

82　Bernard: Father-in-law cited in Kirby, *Evidence of Harm*, 19.

82　Bernard and mercury poisoning: Cited in U.S. Congress, *Mercury in Medicine*, 20.

82　*Medical Hypotheses* paper: S. Bernard, A. Enayati, L. Redwood, H. Roger, and T. Binstock, "Autism: A Novel Form of Mercury Poisoning," *Medical Hypotheses* 56 (2001): 462–71.

83　Safe Minds mission statement: http://www.safeminds.org/home/mission_strategies.html.

83　Redwood and challenge to public health officials: L. Redwood, "Mercury and Autism: Coincidence or Cause and Effect?" *Autism-Asperger's Digest*, July–August 2000.

84　*New York Times* article: A. Allen, "The Not-So-Crackpot Autism Theory," *New York Times Magazine*, November 10, 2002.

84　Arthur Allen on Halsey: A. Allen, interview, November 13, 2007.

84　Geiers' first VAERS paper: M. R. Geier and D. A. Geier, "Neurodevelopmental Disorders After Thimerosal-Containing Vaccines: A Brief Communication," *Experimental Biology and Medicine* 228 (2003): 660–64.

85　Geiers' heart disease paper: M. R. Geier and D. A. Geier, "Thimerosal in Childhood Vaccines, Neurodevelopmental Disorders, and Heart Disease in the United States," *Journal of American Physicians and Surgeons* 8 (2003): 6–10.

85　Geiers' first chelation paper: J. Bradstreet, D. A. Geier, J. J. Kartzinel, J. B. Adams, and M. R. Geier, "A Case-Control Study of Mercury Burden in

Children with Autistic Spectrum Disorders," *Journal of American Physicians and Surgeons* 8 (2003): 76–79.

86 The Hanleys: Cited in S. Kleffman, "Autism Treatment Sparks Hope for Parents," *Knight Ridder Newspapers*, June 10, 2005.

86 Generation Rescue: Mission statement, http://www.generationrescue.org/background.

87 Handley supports Geiers: http://www.generationrescue.org/hall_fame.

87 Baron-Cohen article: S. Baron-Cohen, R. C. Knickmeyer, and M. K. Belmonte, "Sex Differences in the Brain: Implications for Explaining Autism," *Science* 310 (2005): 819–23.

88 Geiers' first Lupron paper: D. A. Geier and M. R. Geier. "The Biochemical Basis and Treatment of Autism: Interactions Between Mercury, Transsulfuration, and Androgens" [submitted to *Autoimmunity Reviews* but not published].

88 Geier on Radio Liberty: Cited in K. Seidel. "Significant Misrepresentations: Mark Geier, David Geier, and the Evolution of the Lupron Protocol," http://www.neurodiversity.com/weblog/article/106/desperation-time.

89 Haley on tubulin: B. E. Haley, "Toxic Overload: Assessing the Role of Mercury in Autism," *Mothering Magazine*, November/December 2002.

89 Redwood praises Haley: Cited in A. Mead and J. Warren, "U.K. Chemist Tilts at Autism's Origins," *Lexington Herald-Leader*, July 24, 2005.

89 Deth paper: M. Waly, H. Olteanu, R. Banerjee et al., "Activation of Methionine Synthase by Insulin-Like Growth Factor-1 and Dopamine: A Target for Neurodevelopmental Toxins and Thimerosal," *Molecular Psychiatry* 9 (2004): 358–70.

90 Deth testimony: Cited in U.S. Congress, House of Representatives, Committee on Government Reform, *Truth Revealed: New Scientific Discoveries Regarding Mercury in Medicine and Autism*, 108th Congress, Second session, September 8, 2004 (Washington, DC: U.S. Government Printing Office, 2004).

90 Hornig study: M. Hornig, D. Chian, and W. I. Lipkin, "Neurotoxic Effects of Postnatal Thimerosal Are Mouse Strain Dependent," *Molecular Psychiatry* 9 (2004): 833–45.

90 Hornig and Attkisson: Kirby, *Evidence of Harm*, 366.

90 Hornig and WebMD: Ibid.

90 Hornig on television: WNYW-TV (Fox), Channel 5, February 14, 2005.

91 Simpsonwood: Quotes from Walter Orenstein, Thomas Verstraeten, Paul Stehr-Green, and Robert Davis were obtained from a transcript of the Simpsonwood meeting, "Scientific Review of Vaccine Safety DataLink Information," Simpsonwood Retreat Center, Norcross, Georgia, June 7–8, 2000.

92 Verstraeten paper: T. Verstraeten, R. L. Davis, F. DeStefano et al., "Study of Thimerosal-Containing Vaccines: A Two-Phased Study of Computerized Health Maintenance Organization Databases," *Pediatrics* 112 (2003): 1039–48.

94   Kennedy and *Rolling Stone*: All quotes from Kennedy in this and the fol-
     lowing paragraph are from R. F. Kennedy Jr., "Deadly Immunity," *Rolling
     Stone*, June 30, 2005.

97   Kennedy on Imus: "Robert F. Kennedy, Jr., on Vaccines and Autism," *Imus
     in the Morning*, June 20, 2005.

97   Kennedy on MSNBC: *Scarborough Country*, MSNBC, June 21, 2005.

98   Kirby and "definitive answer": Kirby, *Evidence of Harm*, xii.

98   Kirby and "hellish, lost world": Ibid.

98   Kirby and public health officials: Ibid., xiii.

98   Imus on Kirby book: *Imus in the Morning*, April 4, 2005.

98   Imus and chemical companies: *Imus in the Morning*, December 14, 2006.

98   Kirby on MSNBC: *Scarborough Country*, MSNBC, March 2, 2006.

98   Kirby and media attention: *Evidence of Harm* listserv, April 14, 2005,
     http://groups.yahoo.com.group/EOHarm.

99   Bernard and conspiracy: Cited in A. Manning, "Mistrust of Medicine
     Rises with Autism Rate," *USA Today*, July 7, 2005.

99   Burton and conspiracy: Cited in Allen, *Vaccine*, 393.

99   Geier and conspiracy: Cited in Kirby, *Evidence of Harm*, 268–69.

99   Lieberman on Imus: "Vaccine Safety," *Imus in the Morning*, May 31,
     2005.

100 Dodd on Imus: "Thimerosal in Vaccines," *Imus in the Morning*, June 22,
     2005.

100 Kerry on Imus: *Imus in the Morning*, March 20, 2006.

100 Weldon: Cited in Kirby, *Evidence of Harm*, 309.

102 Weldon, Gerberding, and Burns: Cited in ibid., 365.

102 Schwarzenegger ban: M. Levin, "Battle Lines Drawn Over Mercury in
     Shots," *Los Angeles Times*, April 10, 2006.

102 Kirby and declining rates: Cited in "Slouching Toward Truth—Autism
     and Mercury," *Citizen Cain*, November 30, 2005, http://www.citizencain
     .blogspot.com/2005/11/slouching-toward-truth-autism-an_30.html.

103 Humiston to Burton: Cited in U.S. Congress, *Mercury in Medicine*, 207.

103 Haley and truth: B. E. Haley, "Toxic Overload"; also "Prof. Boyd Haley
     Explains Mercury-Autism Link to Kentucky Assembly," Generation Res-
     cue Press Release, October 16, 2004.

103 Geiers and autism epidemic: M. R. Geier and D. A. Geier. "Response to
     Critics on the Adverse Effects of Thimerosal in Childhood Vaccines,"
     Journal of American Physicians and Surgeons 8 (2003): 68–70.

103 Redwood and optimism: "New Data Showing First Drop in 20 Years
     Provides Cautious Hope That Vaccine-Induced Autism on Decline," Safe
     Minds Press Release, April 22, 2004.

103 Kirby and one less arrow: D. Kirby, "Autism, Mercury and the California
     Numbers," *Huffington Post*, http://www.huffingtonpost.com/david-kirby/
     autism-mercury-and-the-c_b_4133.html.

103 Geiers' study on declining rates: D. A. Geier and M. R. Geier, "Early
     Downward Trends in Neurodevelopmental Disorders Following Removal

of Thimerosal-Containing Vaccines," *Journal of American Physicians and Surgeons* 11 (2006): 8–13.

104 Deth response to Geiers' study: Cited in "Article on Falling Autism Rates Sparks Controversy," March 6, 2006, http://www.aapsoline.org/nod/newsofday265.php.

104 *Evidence of Harm* movie: Described on Participant Productions Web site, August 17, 2006, http://participantproductions.com/films/In+Development/249/EvidenceofHarm.

104 Kirby response to movie: D. Kirby, *Evidence of Harm* listserv, December 11, 2006 and January 22, 2007.

6. Mercury Falling

106 Stehr-Green study: P. Stehr-Green, P. Tull, M. Stellfeld et al., "Autism and Thimerosal-Containing Vaccines: Lack of Consistent Evidence for an Association," *American Journal of Preventive Medicine* 25 (2005): 101–6.

107 Madsen study: K. M. Madsen, M. B. Lauritsen, C. B. Pedersen et al., "Thimerosal and the Occurrence of Autism: Negative Ecological Evidence from Danish Population-Based Data," *Pediatrics* 112 (2003): 604–6.

107 Hviid study: A. Hviid, M. Stellfeld, J. Wohlfahrt, and M. Melbye, "Association Between Thimerosal-Containing Vaccine and Autism," *Journal of the American Medical Association* 290 (2003): 1763–66.

107 Heron study: J. Heron and J. Golding, "Thimerosal Exposure in Infants and Developmental Disorders: A Prospective Cohort Study in the United Kingdom Does Not Support a Causal Association," *Pediatrics* 114 (2004): 577–83.

108 Andrews study: N. Andrews, E. Miller, A. Grant et al., "Thimerosal Exposure in Infants and Developmental Disorders: A Retrospective Cohort Study in the United Kingdom Does Not Support a Causal Association," *Pediatrics* 114 (2004): 584–91.

108 IOM report: Institute of Medicine, *Immunization Safety Review: Vaccines and Autism.*

108 Fombonne study: E. Fombonne, R. Zakarian, A. Bennett et al., "Pervasive Developmental Disorders in Montreal, Quebec, Canada: Prevalence and Links with Immunization," *Pediatrics* 118 (2006): 139–50.

109 Fombonne explanation: Ibid.

109 Thompson study: W. W. Thompson, C. Price, B. Goodson et al., "Early Thimerosal Exposure and Neuropsychological Outcomes at 7 to 10 Years," *New England Journal of Medicine* 357 (2007): 1281–92.

109 Schechter and Grether study: R. Schechter and J. Grether, "Continuing Increases in Autism Reported to California's Development Services System," *Archives of General Psychiatry* 65 (2008): 19–24; see also J. C. Dooren, "Autism Rate Is Still Rising Despite Vaccine Change," *Wall Street Journal*, January 7, 2008; A. Chang, "No Autism Vaccine Link Seen in California Study," Associated Press, January 7, 2008.

109 Fombonne editorial: E. Fombonne, "Thimerosal Disappears but Autism Remains," *Archives of General Psychiatry* 65 (2008): 15–16.

110 Kirby and *Meet the Press*: *Meet the Press*, August 8, 2005.

110 Rotavirus vaccine and intestinal blockage: T. V. Murphy, P. M. Garguillo, M. S. Massoudi et al., "Intussusception Among Infants Given an Oral Rotavirus Vaccine," *New England Journal of Medicine* 344 (2001): 564–72.

111 Measles vaccine and decrease in platelets: R. A. Oski and J. L. Naiman, "Effect of Live Measles Vaccine on the Platelet Count," *New England Journal of Medicine* 275 (1966): 352–56; Institute of Medicine, "Measles and Mumps Vaccine," in *Adverse Events Associated with Childhood Vaccines: Evidence Bearing on Causality* (Washington, DC: National Academy Press, 1994), 130.

111 Swine flu vaccine and Guillain-Barré Syndrome: L. B. Schonberger, D. J. Bregman, J. Z. Sullivan-Bolyai et al., "Guillain-Barré Syndrome Following Vaccination in the National Influenza Immunization Program, United States, 1976–1977," *American Journal of Epidemiology* 110 (1979): 105–23.

112 SV40 and cancer: A. J. Girardi, B. H. Sweet, V. B. Slotnick, and M. R. Hilleman, "Development of Tumors in Hamsters Inoculated in the Neonatal Period with Vacuolating Virus, SV40," *Proceedings of the Society for Experimental Biology and Medicine* 109 (1962): 649–60; K. V. Shah, "Simian Virus 40 and Human Disease," *Journal of Infectious Diseases* 190 (2004): 2061–64.

112 Adenovirus and cancer: A. J. Girardi, M. R. Hilleman, and R. E. Zwickey, "Tests in Hamsters for Oncogenic Quality of Ordinary Viruses Including Adenovirus Type 7," *Proceedings of the Society for Experimental Biology and Medicine* 115 (1964): 1141–50; V. M. Larson, A. J. Girardi, M. R. Hilleman, and R. E. Zwickey, "Studies of Oncogenicity of Adenovirus Type 7 Viruses in Hamsters," *Proceedings of the Society for Experimental Biology and Medicine* 118 (1965): 15–24.

112 Formaldehyde and cancer: V. S. Goldmacher and W. G. Thilly, "Formaldehyde Is Mutagenic for Cultured Human Cells," *Mutation Research* 116 (1983): 417–22; D. L. Ragan and C. J. Boreiko, "Initiation of C3H/10T1/2 Cell Transformation by Formaldehyde," *Cancer Letters* 13 (1981): 325–31; "Epidemiology of Chronic Occupational Exposure to Formaldehyde: Report of the Ad Hoc Panel on Health Aspects of Formaldehyde," *Toxicology and Industrial Health* 4 (1988): 77–90.

112 The Cutter Incident: Offit, *The Cutter Incident*.

113 Edwin Lennette: *Gottsdanker v. Cutter Laboratories*, District Court of Appeals of the State of California, First Appellate District, 1 Civ. 18413 and 18414, November 20, 1957 to January 31, 1958.

113 Peggy Johnston: Cited in J. Cohen, "Did Merck's Failed HIV Vaccine Cause Harm?" *Science* 318 (2007): 1048–49.

113 Cigarette smoking and lung cancer studies: E. L. Wynder and E. A. Graham, "Tobacco Smoking as a Possible Etiologic Factor in Bronchogenic

Carcinoma: A Study of 684 Proved Cases," *Journal of the American Medical Association* 143 (1950): 334; R. Doll and A. B. Hill, "The Mortality of Doctors in Relation to Their Smoking Habits: A Preliminary Report," *British Medical Journal* 228 (1954): 1451–55.

113 Bradford Hill: A. B. Hill, "Observations and Experiment," *New England Journal of Medicine* 248 (1953): 1000.

114 Mercury: P. W. Davidson, G. J. Myers, and B. Weiss, "Mercury Exposure and Child Development Outcomes," *Pediatrics* 113 (2004): 1023–29; D. O. Marsh, T. W. Clarkson, C. Cox et al., "Fetal Methylmercury Poisoning: Relationship Between Concentration in Single Strands of Maternal Hair and Child Effects," *Archives of Neurology* 44 (1987): 1017–22; L. Magos, A. W. Brown, S. Sparrow et al., "The Comparative Toxicology of Ethyl- and Methyl-Mercury," *Archives of Toxicology* 57 (1985): 260–67; L. Magos, "Review on the Toxicity of Ethylmercury, Including Its Presence as a Preservative in Biological and Pharmaceutical Products," *Journal of Applied Toxicology* 21 (2001): 1–5; P. W. Davidson, G. J. Myers, C. Cox et al., "Effects of Prenatal and Postnatal Methylmercury Exposure from Fish Consumption on Neurodevelopment: Outcomes at 66 Months of Age in the Seychelles Child Development Study," *Journal of the American Medical Association* 280 (1998): 701–7; M. Harada, "Minamata Disease: Methylmercury Poisoning in Japan Caused by Environmental Pollution," *Critical Reviews in Toxicology* 25 (1995): 1–24; P. W. Davidson, D. Palumbo, G. J. Myers et al., "Neurodevelopmental Outcomes of Seychellois Children from the Pilot Cohort at 108 Months Following Prenatal Exposure to Methylmercury from a Maternal Fish Diet," *Environmental Research Section A* 84 (2000): 1–11; F. Bakir, S. F. Damliji, L. Amin-Zaki et al., "Methylmercury Poisoning in Iraq," *Science* 181 (1973): 230–41; T. W. Clarkson, L. Magos, and G. J. Myers, "The Toxicology of Mercury—Current Exposures and Clinical Manifestations," *New England Journal of Medicine* 349 (2003): 1731–37; L. I. Sweet and J. T. Zelikoff, "Toxicology and Immunotoxicology of Mercury: A Comparative Review in Fish and Humans," *Journal of Toxicology and Environmental Health* 4 (2001): 161–205; T. Barkay and I. Wagner-Döbler, "Microbial Transformations of Mercury: Potentials, Challenges, and Achievements in Controlling Mercury Toxicity in the Environment," *Advances in Applied Microbiology* 57 (2005): 1–52; L. Magos and T. W. Clarkson, "Overview of the Clinical Toxicity of Mercury," *Annals of Clinical Biochemistry* 43 (2006): 257–68; A. M. Brownawell, S. Berent, R. L. Brent et al., "The Potential Adverse Health Effects of Dental Amalgam," *Toxicology Reviews* 24 (2005): 1–10; D. W. Boening, "Ecological Effects, Transport, and Fate of Mercury: A General Review," *Chemosphere* 40 (2000): 1335–51; J. F. Risher, C. T. De Rosa, D. E. Jones, and H. E. Murray, "Summary Report for the Expert Panel of the Toxicological Profile for Mercury," *Toxicology and Industrial Health* 15 (1999): 483–516; N. L.

Brown, Y. C. Shih, C. Leang et al., "Mercury Transport and Resistance," *Biochemical Society Transactions* 30 (2002): 715–18; S. A. Counter and L. H. Buchanan, "Mercury Exposure in Children: A Review," *Toxicology and Applied Pharmacology* 198 (2004): 209–30; T. W. Clarkson, "The Three Modern Faces of Mercury," *Environmental Health Perspectives* 110 (2002): 11–23; L. Magos, "Physiology and Toxicology of Mercury," *Metal Ions in Biological Systems* 34 (1997): 321–70; M. Bigham, R. Copes, and L. Srour, "Exposure to Thimerosal in Vaccines Used in Canadian Infant Immunization Programs with Respect to Risk of Neurodevelopmental Disorders," *Canada Communicable Disease Report* 28 (2002): 69–80; M. E. Pichichero, E. Cernichiari, J. Lopreiato, and J. Treanor, "Mercury Concentrations and Metabolism in Infants Receiving Vaccines Containing Thimerosal: A Descriptive Study," *Lancet* 360 (2002): 1737–41; T. M. Burbacher, D. D. Shen, N. Liberato et al., "Comparison of Blood and Brain Mercury Levels in Infant Monkeys Exposed to Methylmercury or Vaccines Containing Thimerosal," *Environmental Health Perspectives* 113 (2005): 1015–21.

115 Nelson and Bauman study: K. B. Nelson and M. L. Bauman, "Thimerosal and Autism?" *Pediatrics* 111 (2003): 674–79.

116 *New York Times* advertisement: *New York Times*, June 8, 2005, A13.

116 *USA Today* advertisement: *USA Today*, April 6, 2006, 16A.

117 Sarah Parker: S. Parker, interview, November 6, 2007.

117 Marie McCormick: M. McCormick, interview, November 13, 2007.

118 Strassel editorial in *Wall Street Journal*: "The Politics of Autism: Lawsuits and Emotion vs. Childhood Vaccines," *Wall Street Journal*, December 29, 2003.

118 Kim Strassel: K. Strassel, interview, November 15, 2007.

119 CDC threats: Cited in G. Harris and A. O'Connor, "On Autism's Causes, It's Parents vs. Research," *New York Times*, June 25, 2005.

119 Melinda Wharton: Ibid.

119 Salisbury: D. Salisbury, interview, January 9, 2008.

120 Stephen Edelson: Cited in U.S. Congress, House of Representatives, Committee on Government Reform, *The Future Challenges of Autism: A Survey of the Ongoing Initiatives in the Federal Government to Address the Epidemic*, 108th Congress, First session, November 20, 2003 (Washington, DC: U.S. Government Printing Office, 2004), 137.

120 Paul Harch: Cited in U.S. Congress, House of Representatives, Committee on Government Reform, *Autism Spectrum Disorders: An Update of Federal Government Initiatives and Revolutionary New Treatment of Neurodevelopmental Diseases*, 108th Congress, Second session, May 6, 2004 (Washington, DC: U.S. Government Printing Office, 2004), 61.

121 Controversial therapies: Details supplied by Camille Clark, personal communication, October 14, 2007. Clark is discussed in chapter 11.

121 DAN therapies: Autism Research Institute, DAN Conference Proceedings, Seattle, WA, October 6–9, 2006.

122 Reichelt study: K. L. Reichelt, K. Hole, and A. Hamberger, "Biologically
    Active Peptide-Containing Fractions in Schizophrenia and Childhood Au-
    tism," *Advances in Biochemical Psychopharmacology* 28 (1993): 627–43.

122 Refutation of Reichelt study: A. LeCouteur, O. Trygstad, C. Evered et al.,
    "Infantile Autism and Urinary Excretion of Peptides and Protein-Associated
    Peptide Complexes," *Journal of Autism and Developmental Disorders* 18
    (1988): 181–90; L. Pavone, A. Fiumara, G. Bottaro et al., "Autism in Ce-
    liac Disease: Failure to Validate the Hypothesis That a Link Might Exist,"
    Biological Psychiatry 42 (1997): 72–75; L. C. Hunter, A. O'Hare, W. J.
    Herron et al., "Opioid Peptides and Dipeptidyl Peptidase in Autism," *De-
    velopmental Medicine and Child Neurology* 45 (2003): 121–28.

122 Elimination diets and bone thinning: "Thin Bones Seen in Boys with Au-
    tism and Autism Spectrum Disorder," press release, National Institutes
    of Health, January 29, 2008, http://www.nih.gov/news/health/jan2008/
    nichd-29.htm.

123 Risks of alternative therapies: Camille Clark, personal communication,
    October 14, 2007.

123 North Carolina secretin study: A. D. Sandler, K. A. Sutton, J. DeWeese et al.,
    "Lack of Benefit of a Single Dose of Synthetic Human Secretin in the
    Treatment of Autism and Pervasive Developmental Disorder," *New En-
    gland Journal of Medicine* 341 (1999): 1801–6.

124 Humiston: Cited in U.S. Congress, House of Representatives, Committee
    on Government Reform, *Mercury in Medicine: Are We Taking Unneces-
    sary Risks?* 106th Congress, Second session, July 18, 2000 (Washington,
    DC: U.S. Government Printing Office, 2001), 207.

124 Laidler: J. R. Laidler, "Through the Looking Glass: My Involvement with
    Autism Quackery," http://www.autism-watch.org/about/bio2.shtml.

126 California autism rates: Schechter and Grether, "Continuing Increases in
    Autism."

126 Robert Davis: Cited in Kirby, *Evidence of Harm*, 276.

127 Arthur Allen on new data: A. Allen, interview, November 13, 2007; "The
    Not-So-Crackpot Autism Theory," *New York Times Magazine*, Novem-
    ber 10, 2002.

127 Arthur Allen on debate: A. Allen interview, November 13, 2007.

128 Seidel on reason for involvement: K. Seidel, interview, January 24, 2008.

128 Seidel on letter to Kirby: Ibid.

128 Seidel letter to Kirby: K. Seidel, May 29, 2005.

7. Behind the Mercury Curtain

Kathleen Seidel's background and quotes were obtained from interviews on
November 5, 2007, and January 24, 2008.

132 Haley as expert witness: "Opinion of the Court in the RhoGAM Trial,"
    http://www.kevinleitch.co.uk/wp/?p=393.

132 Haley in court, 2008: *Pamela and Ernest Blackwell v. Sigma Aldrich, Inc.,* Circuit Court for Baltimore City, Case Number 24-C-04-004829.

133 Haley accusations: Ibid.

133 Haley and chelators: Dr. Neubrander Bio Chat, http://drneubrander.com.

133 McCormick on Haley: Cited in A. Mead and J. Warren, "U.K. Chemist Tilts at Autism's Origins," *Lexington Herald-Leader,* July 24, 2005.

133 Haley on McCormick: Cited in ibid.

133 NICO: J. E. Dodes and S. Barrett, "A Critical Look at Cavitational Osteopathosis, NICO, and 'Biological Dentistry,'" *Quackwatch,* http://www.quackwatch.org/01QuackeryRelatedTopics/cavitation.html.

134 Mark Geier and in vitro fertilization: Cited in C. Lowallen, "Breeding the Perfect Human," *Sunday Mail,* October 9, 1988.

134 Mark Geier and DPT vaccine: T. Tyler, "Tests on Vaccine Are Inadequate, Geneticist Says," *Toronto Star,* December 3, 1987.

135 Mark and David Geier's home laboratory: G. Harris and A. O'Connor, "On Autism's Causes, It's Parents vs. Research," *New York Times,* June 25, 2005.

136 Mark Geier in court: Reviewed in *Weiss v. Secretary of the Department of Health and Human Services,* filed October 9, 2003, case number 03-190V.

136 Geiers' MMR paper: M. R. Geier and D. A. Geier, "Pediatrics MMR Vaccination Safety," *International Pediatrics* 18 (2003): 108–13.

137 Geiers' thimerosal paper: M. R. Geier and D. A. Geier, "Thimerosal in Childhood Vaccines, Neurodevelopmental Disorders, and Heart Disease in the United States," *Journal of American Physicians and Surgeons* 8 (2003): 6–10.

137 Flaws in Geiers' study: "Study Fails to Show a Connection Between Thimerosal and Autism," American Academy of Pediatrics, May 16, 2003, http://www.aap.org/profed/thimaut-may03.htm.

137 VAERS data supplied by personal-injury lawyers: M. J. Goodman and J. Nordin, "Vaccine Adverse Event Reporting System Reporting Source: A Possible Source of Bias in Longitudinal Studies," *Pediatrics* 117 (2006): 387–90.

138 Orient on *Nightline:* "Vaccines and Their Risks," *Nightline,* October 14, 1999.

138 Pickering on epidemiological studies: L. Pickering, interview, October 31, 2007.

138 Criteria study: S. K. Parker, B. Schwartz, J. Todd, and L. K. Pickering, "Thimerosal-Containing Vaccines and Autistic Spectrum Disorder: A Critical Review of Published Original Data," *Pediatrics* 114 (2004): 793–804.

138 Pickering on eight criteria: L. Pickering, interview, October 31, 2007.

138 Pickering and lawsuit: Ibid.

139 Geiers' lawsuit: *Dr. Mark Geier and David Geier vs. Department of Health and Human Services, Dr. Sarah K. Parker, Dr. James Todd, Dr. Benjamin Schwartz, Dr. Larry Pickering, and the American Academy of Pediatrics,* filed September 1, 2005.

139 Pickering and lawsuit: L. Pickering, interview, October 31, 2007.

139 Parker and lawsuit: S. Parker, interview, November 6, 2007.

139 Pickering and lawsuit: L. Pickering, interview, October 31, 2007.

139 Geier testimony in court: *Easter v. American Home Products*, filed December 22, 2004, case number 1:2004mc00586.

140 David Geier at Autism One conference: Transcript of D. Geier talk and accompanying PowerPoint presentation found in "Significant Misrepresentations: Mark Geier, David Geier and the Evolution of the Lupron Protocol (Part Seven): An Unnatural Bond," July 24, 2007, http://www.neurodiversity.com.

141 Geier and Lupron: Cited in "Significant Misrepresentations: Mark Geier, David Geier and the Evolution of the Lupron Protocol (Part Six): Desperation Time," July 20, 2007, http://www.neurodiversity.com; emphasis added.

141 Seidel and IRB: Office for Human Research Protection, U.S. Department of Health and Human Services, IRB Number IRB00005375, found in "Significant Misrepresentations: Mark Geier, David Geier and the Evolution of the Lupron Protocol (Part Two): An Elusive Institute," June 20, 2006, http://www.neurodiversity.com.

142 Shoemaker on Mark Geier: Cited in "Significant Misrepresentations: Mark Geier, David Geier and the Evolution of the Lupron Protocol (Part Five): Testimony of the Faithful," July 12, 2006, http://www.neurodiversity.com.

143 Seidel letter on use of trade names: K. Seidel, letter to Dr. Shoenfeld, editor-in-chief of *Autoimmunity Reviews*, November 10, 2006.

143 Geiers' patents: "Methods for Screening, Studying, and Treating Disorders with a Component of Mercurial Toxicity," file number 20060058271, filed September 16, 2004; "Methods of Treating Disorders Having a Component of Mercury Toxicity," file number 20060058241, filed September 15, 2005.

144 TAP prosecution by federal government: "Significant Misrepresentations: Mark Geier, David Geier and the Evolution of the Lupron Protocol (Part Ten): TAP's Connection," August 29, 2006, http://www.neurodiversity.com; S. P. Duffy, "Pharmaceutical Companies to Pay $1.2B in Medicare Fraud," *Legal Intelligencer*, June 24, 2003; "TAP Pharmaceutical Products Inc. and Seven Others Charged with Health Care Crimes: Company Agrees to Pay $875 Million to Settle Charges," press release, U.S. Department of Justice, October 3, 2001; D. Gellene, "Drug Company Agrees to Plead Guilty: Will Pay Out $875 Million Penalty," *Los Angeles Times*, October 5, 2001; J. Strax, "Lupron Kickbacks Betrayed Prostate Cancer Patient Trust," *PSA Rising*, February 27, 2001; J. Appleby, "TAP Pharmaceuticals Fined $875 Million," *USA Today*, October 3, 2001.

144 Seidel on culture that allows Geiers to flourish: "Significant Misrepresentations: Mark Geier, David Geier and the Evolution of the Lupron Protocol (Part Five): Testimony of the Faithful," July 12, 2006, http://www.neurodiversity.com.

145 Handley and neurodiversity bloggers: E-mail from J. B. Handley to http://neurodiversity.com, October 11, 2005.

145 IOM and chelation: Institute of Medicine, *Immunization Safety Review: Thimerosal-Containing Vaccines and Neurodevelopmental Disorders*, 47.

146 Tariq Nadama death: "Deaths Associated with Hypocalcemia from Chelation Therapy—Texas, Pennsylvania, and Oregon, 2003–2005," *Morbidity and Mortality Weekly Report* 55 (2006): 204–7; K. Kane: "Death of 5-Year-Old Boy Linked to Controversial Chelation Therapy," *Pittsburgh Post-Gazette*, January 6, 2006; "Drug Error, Not Chelation Therapy, Killed Boy, Experts Say," *Pittsburgh Post-Gazette*, January 18, 2006; V. Linn, "Parents of Children with Autism Discuss Results of Chelation," *Pittsburgh Post-Gazette*, August 29, 2005; J. C. Yates, "Autistic Boy's Death Raises Questions About Medical Treatment," Associated Press, August 26, 2005; K. Kane and V. Linn, "Boy Dies During Autism Treatment," *Pittsburgh Post-Gazette*, August 25, 2006; M. Fitzpatrick, "When Quackery Kills," http://www.spiked-online.com, November 4, 2005; "PA Files Disciplinary Charges in Autistic Boy's Death," Associated Press, October 31, 2006; B. C. Rittmeyer, "Police Raid Mercer County Doctor's Office," *Pittsburgh Tribune-Review*, April 12, 2007; J. Mandak, "Doctor Charged in Autistic Boy's Death," Associated Press, August 22, 2007; R. Plushnick-Masti, "Doctor to Trial in Autistic Boy's Death," Associated Press, November 16, 2007; K. Kane, "Doctor Who Used Chelation Therapy Charged in Autistic Boy's Death," *Pittsburgh Post-Gazette*, August 23, 2007.

147 Rescue Angel on chelation: Cited in Linn, "Parents of Children with Autism Discuss Results of Chelation."

147 Gismondi on criminal charges: Cited in Mandak, "Doctor Charged in Autistic Boy's Death."

147 Deth study: M. Waly, H. Olteanu, R. Banerjee et al., "Activation of Methionine Synthase by Insulin-Like Growth Factor-1 and Dopamine: A Target for Neurodevelopmental Toxins and Thimerosal," *Molecular Psychiatry* 9 (2004): 358–70.

148 Deth supported by Safe Minds: "Form 990 Tax Return," Safe Minds, 2003, employer number 22:3767992.

148 Hornig study: M. Hornig, D. Chian, and W. I. Lipkin, "Neurotoxic Effects of Postnatal Thimerosal Are Mouse Strain Dependent," *Molecular Psychiatry* 9 (2004): 833–45.

148 IOM on Hornig study: Institute of Medicine. *Immunization Safety Review: Vaccines and Autism*, 138.

148 Failure to reproduce Hornig study: R. F. Berman, I. N. Pessah, P. R. Mouton et al., "Low-Level Neonatal Thimerosal Exposure: Further Evidence of Altered Neurotoxic Potential in SJL Mice," *Toxicological Sciences* 101 (2008): 294–309.

148 Hornig supported by Safe Minds: "Form 990 Tax Return," Safe Minds, 2003, employer number 22:3767992.

149 Robert F. Kennedy Jr. on Imus: "Robert F. Kennedy, Jr., on Vaccines and Autism," *Imus in the Morning*, June 20, 2005.

149 Robert F. Kennedy Jr. arrest: "Robert Kennedy Jr. Admits He Is Guilty in Possessing Heroin," *New York Times*, February 18, 1984.

149 Walter Olson regarding Robert F. Kennedy Jr.: W. Olson, interview, October 17, 2007.

149 Robert F. Kennedy Jr. and pork producers: Discover the Networks.Org, http://www.discoverthenetworks.org/individualProfile.asp?indid=693.

150 Kennedy and Levin Papantonio: W. Olson, interview, October 17, 2007.

150 Olson and *Rolling Stone*: Ibid.

150 Kirby and interviews for *Evidence of Harm*: Kirby, *Evidence of Harm*, xiii.

151 Curtis Allen and Kirby: C. Allen, interview, October 19, 2007.

151 Kirby and journalistic neutrality: E. Arranga, D. Kirby, and T. Small, "Interview with David Kirby Concerning 'Evidence of Harm' Associated with Thimerosal-Containing Vaccines," *Medical Veritas* 2 (2005): 1–9.

151 Arthur Allen and Kirby: A. Allen, interview, November 13, 2007.

152 Arthur Allen and Kirby's book: Ibid.

152 Kirby and alternative points of view: Kirby, *Evidence of Harm*, xiii.

153 Kirby and NPR: Cited in "NPR, Air America Nix Autism/Vaccine Book," *Corporate Crime Reporter*, April 21, 2005.

153 Kirby laments Imus firing: D. Kirby, "Imus, Autism, and America," *Huffington Post*, April 14, 2007.

153 Waxman and scientists: U.S. Congress, *Autism: Present Challenges, Future Needs*, 190.

154 Redwood lawsuit: Lyndelle and William Thomas Redwood on behalf of Will Redwood, Atlanta, Georgia, Court of Federal Claims number 04-0402V.

154 Burton and pharmaceutical companies: U.S. Congress, House of Representatives, Committee on Government Reform, *Truth Revealed: New Scientific Discoveries Regarding Mercury in Medicine and Autism*, 108th Congress, Second session, September 8, 2004 (Washington, DC: U.S. Government Printing Office, 2004), 201.

154 Kirby and pharmaceutical companies: "Evidence of Harm Follow-Up," *Imus in the Morning*, April 4, 2005.

155 Personal-injury lawyer predicts thimerosal litigation: Cited in B. Rimland, "The Autism-Vaccine Disaster," *Autism Research Review International* 16 (2002): 3.

8. Science in Court

Unless otherwise stated, all quotes from George Hastings, Thomas Powers, Vincent Matanoski, Sylvia Chin-Caplan, Theresa Cedillo, Vas Aposhian, Arthur Krigsman, Vera Byers, Marcel Kinsbourne, Eric Fombonne, Stephen Bustin, and Nicholas Chadwick were taken from the Omnibus Autism Proceeding,

Federal Claims Court, Washington, D.C., http://www.uscfc.uscourts.gov/omni bus-autism-proceeding.

156 National Childhood Vaccine Injury Act: G. Evans, D. Harris, and E. M. Levine, "Legal Issues," in Plotkin, Orenstein, and Offit, *Vaccines*; T. Mauro, "Vaccine Test Case Reaches Federal Court," *Legal Times*, June 4, 2007.

157 Bendectin: The two best sources for information on Bendectin are Green, *Bendectin and Birth Defects*, and Huber, *Galileo's Revenge*, 111–29.

158 Kevin Conway on Omnibus Autism Proceeding: Cited in Mauro, "Vaccine Test Case Reaches Federal Court."

158 Federal judges in Omnibus Autism Proceeding: M. Fox, "Washington Court Will Hear Autism-Vaccine Suits," http://www.reuters.com, June 10, 2007.

162 Michelle Cedillo in court: G. Harris, "Opening Statements in Case on Autism and Vaccinations," *New York Times*, June 12, 2007.

162 Theresa and Michael Cedillo: B. Meadows and S. Mandel, "Vaccinations on Trial," *People*, July 2, 2007.

170 Bustin: Cited in M. Fitzpatrick, "The MMR-Autism Theory? There's Nothing in It," http://www.spiked-online.com, July 4, 2007.

174 Andrew Wakefield and David Brown: D. Salisbury, interview, October 24, 2007.

9. Science and the Media

177 *Meet the Press*: Tim Russert, interview with Harvey Fineberg and David Kirby, *Meet the Press*, August 7, 2005.

178 Harvey Fineberg: H. Fineberg, interview, November 1, 2007.

179 Judea Pearl: J. Pearl, "The Daniel Pearl Standard," *Wall Street Journal*, January 30, 2008.

180 Flat Earth Society: http://www.alaska.net/~clund/e_djublonskopf/Flat earthsociety.htm.

181 Arthur Allen: A. Allen, interview, November 13, 2007.

182 Harris and O'Connor *New York Times* article: G. Harris and A. O'Connor, "On Autism's Causes, It's Parents vs. Research," *New York Times*, June 25, 2005.

182 Daniel Schulman: D. Schulman, "Drug Test," *Columbia Journalism Review*, November/December 2005.

182 Blogger to Westover: Cited in ibid.

183 Westover to blogger: Cited in ibid.

183 Pinker and apple cart: S. Pinker, review of N. Angier, "The Canon: A Whirligig Tour of the Beautiful Basics of Science," *New York Times Book Review*, May 27, 2007.

184 Phillips on Wakefield: M. Phillips, "MMR: The Truth," *Daily Mail*, March 11, 2003.

184 Johnson on science as subversive: G. Johnson, review of F. Dyson, "The Scientist as Rebel," *New York Times Book Review*, January 7, 2007.

184 Haldane on scientists and reason: Cited in ibid.

186 Shermer on lone scientists: Shermer, *Why People Believe Weird Things*, 50.

186 Relman on publishing science: Cited in Milloy, *Junk Science Judo*, 170.

186 MacMahon study: B. MacMahon, S. Yen, D. Trichopoulos et al., "Coffee and Cancer of the Pancreas," *New England Journal of Medicine* 304 (1981): 630–33.

187 Shermer on fallibility of science: Shermer, *Why People Believe Weird Things*, 21.

187 Park and standing on loose soil: Park, *Voodoo Science*, 39.

188 Hill and Knowlton: Hill and Knowlton's influence and all quotes on the cigarette smoking controversy can be found in Brandt, *The Cigarette Century*.

191 Russell's teapot analogy: B. Russell, "Is There a God?" http://evans-experientialism.freewebspace.com/russell10.htm.

191 Berger opinion: *Pamela and Ernest Blackwell v. Sigma Aldrich, Inc.*, Circuit Court for Baltimore City, case number 24-C-04-004829, 2008.

193 Park on confronting voodoo science: Park, *Voodoo Science*, 27.

194 Fombonne and beliefs: Omnibus Autism Proceeding, http://www.uscfc.uscourts.gov/omnibus-autism-proceeding.

195 Park on eminent scientists: Park, *Voodoo Science*, 9.

10. Science and Society

197 Church ruling against Galileo: "The Crime of Galileo: Indictment and Abjuration of 1633," http://www.fordham.edu/halsall/mod/1630galileo.html.

197 Burton on CNN: *Talk Back Live*, CNN, October 5, 1999.

197 Burton and potential conflicts: U.S. Congress, *Autism: Present Challenges, Future Needs*, 194.

198 Milloy on suppressing science: Milloy and Gough, *Silencing Science*, vi.

198 Breast-implant litigant on scientific funding: Cited in ibid., 37.

198 Haley on pharmaceutical companies: Cited in A. Mead and J. Warren, "U.K. Chemist Tilts at Autism's Origins," *Lexington Herald-Leader*, July 24, 2005.

199 Unethical acts by pharmaceutical companies: G. Harris: "Abbott to Pay $622 Million to End Inquiry Into Marketing," *New York Times*, June 27, 2003; "Medical Marketing Treatment by Incentive; As Doctor Writes Prescription, Drug Company Writes a Check," *New York Times*, June 27, 2004; "Guilty Pleas Seen for Drug Maker," *New York Times*, July 16, 2004; "FDA Seizes Millions of Pills from Pharmaceutical Plants," *New York Times*, March 5, 2005.

199 Rothman and new McCarthyism: K. Rothman, "Conflict of Interest: The New McCarthyism in Science," *Journal of the American Medical Association* 269 (1993): 2782–84.

200 Horton on Wakefield: T. Moynihan, "Lancet Doubts Validity of Controversial MMR Report," *Press Association*, February 20, 2004.

201 Cold fusion: Taubes, *Bad Science*.

202 University of Google: Jenny McCarthy on the *Oprah Winfrey Show*, September 18, 2007.

202 Muir Gray on postmodernism: J. A. Muir Gray, "Post-Modern Medicine," *Lancet* 354 (1999): 1550–53.

203 Waxman and scientific expertise: U.S. Congress, *Autism: Present Challenges, Future Needs*, 190.

204 Smith on postmodernism: R. Smith, "The Discomfort of Patient Power," *British Medical Journal* 324 (2002): 497–98.

204 Smith on ignoring medical advice: Ibid.

204 Fitzpatrick on respecting expertise: Fitzpatrick, *MMR and Autism*, 187.

204 Fitzpatrick on pragmatism: Ibid.

205 Birth of NCCAM: J. Groopman, "No Alternative," *Wall Street Journal*, August 7, 2006.

206 Alternative medicines: Ibid.

207 Groopman on distinguishing magic from medicine: Ibid.

207 Pratt at Burton hearing: Cited in U.S. Congress, *Autism: Present Challenges, Future Needs*, 441.

207 Park on goals of a civilized society: Park, *Voodoo Science*, 211.

207 Strauss on anecdotes: Cited in Groopman, "No Alternative."

209 IOM report on MMR: Institute of Medicine, *Immunization Safety Review: Measles-Mumps-Rubella Vaccine and Autism*.

209 Burton on Institute of Medicine report: Cited in B. Vastag, "Congressional Autism Hearings Continue: No Evidence MMR Vaccine Causes Disorder," *Journal of the American Medical Association* 285 (2001): 2567–69.

209 Incidence of autism in 1998: E. Sponheim and O. Skjeldal, "Autism and Related Disorders: Epidemiological Findings in a Norwegian Study Using ICD-10 Diagnostic Criteria," *Journal of Autism and Developmental Disorders* 28 (1998): 217–27; E. Fombonne, "The Epidemiology of Autism: A Review," *Psychological Medicine* 29 (1999): 769–86.

210 Shermer on making connections: Shermer, *Why People Believe Weird Things*, 7.

210 Park on patterns: Park, *Voodoo Science*, 38.

211 Park on ancient beliefs: Ibid., 9.

211 Popular beliefs: G. H. Gallup Jr. and F. Newport, "Belief in Paranormal Phenomena Among Adult Americans," *Skeptical Inquirer* 15 (1991): 137–47.

212 Wakefield "Through a Glass Darkly" paper: A. J. Wakefield and S. M. Montgomery, "Measles, Mumps, Rubella Vaccine: Through a Glass Darkly," *Adverse Drug Reactions and Toxicological Reviews* 19 (2000): 265–83.

212 Bridget Wakefield: Cited in B. Deer, "MMR: The Truth Behind the Crisis," *Sunday Times* (London), February 22, 2004.

212 Good News Doctor Foundation: Cited in Horton, *MMR Science*, 31.

213 Sykes and deceivers: Cited in "A Plaintiff in the Pulpit: Problems with United Methodist Church Advocacy on Behalf of Vaccine-Injury Litigants,"

letter written by Kathleen Seidel to the Council of Bishops, the General Board of Global Ministries, the General Board of Church and Society, the Women's Division of the United Methodist Church, and the Virginia Interfaith Center for Public Policy, April 12, 2007, http://www.neurodiversity.com/weblog/article/126.

213 Sykes in 2006: Ibid.

213 Sykes in 2007: Ibid.

214 Seidel on symbols: K. Seidel, interview, November 5, 2007.

214 Rescue Angel: Cited in V. Linn, "Parents of Children with Autism Discuss Results of Chelation: Debate Over Controversial Treatment Heats Up After Death of 5-Year-Old Boy," *Pittsburgh Post-Gazette*, August 29, 2005.

214 Halloween candy: Heath and Heath. *Made to Stick*, 14.

216 Milloy and *Dateline NBC*: Milloy, *Junk Science Judo*, 11.

216 Penn and Teller on *YouTube*: http://www.youtube.com/watch?v=yi3erdgVVTw.

216 Fumento on risk: Fumento, *Science Under Siege*.

11. A Place for Autism

Unless otherwise specified, all quotes in this chapter were obtained from interviews with Peter Hotez and Kathleen Seidel on October 26 and November 5, 2007, respectively, and from a personal communication from Camille Clark on October 18, 2007.

218 Genetics: I. Rapin: "Autism," *New England Journal of Medicine* 337 (1997): 97–104, "The Autistic-Spectrum Disorders," *New England Journal of Medicine* 347 (2002): 302–3; J. Piven, "The Biological Basis of Autism," *Current Opinion in Neurobiology* 7 (1997): 708–12; M. M. Bristol, D. J. Cohen, E. J. Costello et al., "State of the Science in Autism: Report of the National Institutes of Health," *Journal of Autism and Developmental Disorders* 26 (1996): 121–54; S. Vedantam, "Autism Risk Rises with Age of Father," *Washington Post*, September 5, 2006; E. Stokstad, "New Hints Into the Biological Basis of Autism," *Science* 294 (2001): 34–37; P. M. Rodier, "The Early Origins of Autism," *Scientific American*, February 2000; P. M. Rodier, J. L. Ingram, B. Tisdale et al., "Embryological Origin for Autism: Developmental Anomalies of the Cranial Nerve Motor Nuclei," *Journal of Comparative Neurology* 379 (1996): 247–61; K. Wong, "The Search for Autism's Roots," *Nature* 411 (2001): 882–84; International Molecular Genetic Study of Autism Consortium, "A Genomewide Screen for Autism: Strong Evidence for Linkage to Chromosomes 2q, 7q, and 16p," *American Journal of Human Genetics* 69 (2001): 570–81; T. H. Wassink, J. Piven, V. J. Vieland et al., "Evidence Supporting WNT2 as an Autism Susceptibility Gene," *American Journal of Medical Genetics* 105 (2001): 406–13; J. L. Ingram, C. J. Stodgell, S. L. Hyman et al., "Discovery of Allelic Variants of HOXA1 and HOXB1: Genetic Sus-

ceptibility to Autism Spectrum Disorders," *Teratology* 62 (2000): 393–405; S. J. Spence, "The Genetics of Autism," *Seminars in Pediatric Neurology* 11 (2004): 196–204; C. Dennis, "All in the Mind of a Mouse," *Nature* 438 (2005): 151–52; J. Sebat, B. Lakshmi, D. Malhotra et al., "Strong Association of De Novo Copy Number Mutations with Autism," *Science Express*, March 15, 2007; "Cause of Autism Narrowed Down to 100 Genes," *All Things Considered*, National Public Radio, March 15, 2007; T. W. Briggs, "Older Parents May Be Risk Factor for Autism," *USA Today*, April 4, 2007; S. M. Klauk, "Genetics of Autism Spectrum Disorder," *European Journal of Human Genetics* 14 (2006): 714–20; T. H. Wassink, L. M. Brzustowicz, C. W. Bartlett, and P. Szatmari, "The Search for Autism Disease Genes," *Mental Retardation and Developmental Disabilities Reviews* 10 (2004): 272–83; R. Muhle, S. V. Trentacoste, and I. Rapin, "The Genetics of Autism," *Pediatrics* 113 (2004): 472–86; M. T. Miller, K. Strömland, L. Ventura et al., "Autism Associated with Conditions Characterized by Developmental Errors in Early Embryogenesis: A Mini Review," *International Journal of Developmental Neuroscience* 23 (2005): 201–19.

218 Cook on genetics: Cited in Omnibus Autism Proceeding, June 17, 2007, http://www.uscfc.uscourts.gov/omnibus-autism-proceeding.

219 2007 study: R. Moessner, C. R. Marshall, J. S. Sutcliffe et al., "Contribution of SHANK-3 Mutations to Autism Spectrum Disorder," *American Journal of Human Genetics* 81 (2007): 1289–97.

219 Home movie studies: J. L. Adrien, A. Perrot, D. Sauvage et al., "Early Symptoms in Autism from Family Home Movies," *Acta Paedopsychiatrica* 55 (1992): 71–75; P. Teitelbaum, O. Teitelbaum, J. Nye et al., "Movement Analysis in Infancy May Be Useful for Early Diagnosis of Autism," *Proceedings of the National Academy of Sciences* 95 (1998): 13982–87; A. E. Mars, J. E. Mauk, and P. W. Dowrick, "Symptoms of Pervasive Developmental Disorders as Observed in Prediagnostic Home Videos of Infants and Toddlers," *Journal of Pediatrics* 132 (1998): 500–504; J. L. Adrien, P. Lenoir, J. Martineau et al., "Blind Ratings of Early Symptoms of Autism Based Upon Family Home Movies," *Journal of the American Academy of Childhood and Adolescent Psychiatry* 32 (1993): 617–26; R. Palomo, M. Belinchón, and S. Ozonoff, "Autism and Family Home Movies: A Comprehensive Review," *Developmental and Behavioral Pediatrics* 27 (2006): 59–68.

219 Canadian study: L. Zwaigenbaum, S. Bryson, T. Rogers et al., "Behavioral Manifestations of Autism in the First Year of Life," *International Journal of Developmental Neuroscience* 23 (2005): 143–52.

219 Thalidomide: K. Strömland, V. Nordin, M. Miller et al., "Autism in Thalidomide Embryopathy: A Population Study," *Developmental Medicine and Child Neurology* 36 (1994): 351–56; Rodier, "The Early Origins of Autism."

219 Rubella: R. B. Feldman, R. Lajoie, L. Mendelson, and L. Pinsky, "Congenital Rubella and Language Disorders," *Lancet* 2 (1971): 978; S. Chess,

P. Fernandez, and S. Korn, "Behavioral Consequences of Congenital Rubella," *Journal of Pediatrics* 93 (1978): 699–703; C. N. Swisher and L. Swisher, "Congenital Rubella and Autistic Behavior," *New England Journal of Medicine* 293 (1975): 198.

220 RhoGam study: J. H. Miles and T. N. Takahashi, "Lack of Association Between Rh Status, Rh Immune Globulin in Pregnancy, and Autism," *American Journal of Medical Genetics* 143 (2007): 1397–1407.

220 Minamata Bay and Iraq: D. O. Marsh, T. W. Clarkson, C. Cox et al., "Fetal Methylmercury Poisoning: Relationship Between Concentration in Single Strands of Maternal Hair and Child Effects," *Archives of Neurology* 44 (1987): 1017–22; P. W. Davidson, G. J. Myers, and B. Weiss, "Mercury Exposure and Child Development Outcomes," *Pediatrics* 113 (2004): 1023–29; M. Harada, "Minamata Disease: Methylmercury Poisoning in Japan Caused by Environmental Pollution," *Critical Reviews in Toxicology* 25 (1995): 1–24.

220 Bourgeron regarding synaptic proteins: Cited in K. Garber, "Autism's Cause May Reside in Abnormalities at the Synapse," *Science* 317 (2007): 190–91.

231 Grinker and hope: Grinker, *Unstrange Minds*, 5.

232 Grinker on research funding: Ibid., 9.

232 Grinker on vaccines: Ibid., 14.

233 Grinker on Isabel: Ibid., 284.

Epilogue

235 Jenny McCarthy on *Oprah*: Interview by Oprah Winfrey, *Oprah Winfrey Show*, NBC, September 18, 2007.

237 Jenny McCarthy on *Good Morning America*: Interview by Diane Sawyer, *Good Morning America*, ABC, September 24, 2007.

237 McCarthy on curing autism: J. McCarthy, *Louder than Words: A Mother's Journey in Healing Autism* (New York: Dutton, 2007), 104.

238 McCarthy and B_{12} shots: Ibid., 139.

238 McCarthy and Diflucan: Ibid., 165.

238 Jenny McCarthy on *Larry King Live*: Interview by Larry King, *Larry King Live*, CNN, September 26, 2007.

239 Larry King on Kartzinel: Ibid.

239 Kartzinel on treating autism: Cited in McCarthy, *Louder than Words*, xvi.

239 Kartzinel and Jenny McCarthy on *Larry King Live*: Interview by Larry King, *Larry King Live*, CNN, September 26, 2007.

241 McCarthy on toxins: McCarthy, *Louder than Words*, 181.

241 Fineberg and offering cures: H. Fineberg, interview, November 1, 2007.

243 *Eli Stone*: Broadcast on ABC, January 31, 2008.

244 Wyatt: E. Wyatt, "ABC Drama Takes on Science and Parents," *New York Times*, January 23, 2008.

245 Articles questioning ABC's judgment: B. Kruskal and C. Allen, "Perpetrating the Autism Myth," *Boston Globe*, January 31, 2008; J. Steir, "ABC's

Autism Outrage," *New York Post*, January 31, 2008; "*Eli Stone*'s Leap of Faith," *New York Times*, February 2, 2008; J. Kornblum, "First-Episode Controversy: The Vaccine-Autism Link," *USA Today*, January 29, 2008.

245 Influenza deaths: S. D. James, "Vaccine-Autism Debate Moves to Small Screen," *ABC News*, http://www.abcnews.com/print?id=4218078.

245 AAP letter to ABC: R. R. Jenkins, letter to Anne Sweeney, January 25, 2008.

245 AAP on canceling *Eli Stone*: "American Academy of Pediatrics Calls for Cancellation of ABC's *Eli Stone* Premiere," AAP press release, January 28, 2008.

245 ABC fails to respond to AAP: James, "Vaccine-Autism Debate Moves to Small Screen."

245 Berlanti: Cited in J. Ivory, eFluxMedia, January 29, 2008, http://www .efluxmedia.com/action-print-n_id-13345.html.

245 Minshew: Cited in M. Roth, "Pitt Expert Goes Public to Counter Fallacy on Autism," *Pittsburgh Post-Gazette*, January 31, 2008.

245 Lawyer defends children: A. M. Schwartz, letter to Anne Sweeney, January 30, 2008.

246 Goldberg on stakes: E. Goldberg, "Is There a Link Between Vaccines and Autism?" *American Lawyer*, August 1, 2007.

246 Sugarman on litigation: Cited in J. Interlundi, "Autism and Vaccines: A Coming Wave of Lawsuits?" *Newsweek*, September 26, 2007.

246 *USA Today* advertisement: *USA Today*, September 25, 2007.

247 Sugarman on jurors: Cited in Interlundi, "Autism and Vaccines."

SELECTED BIBLIOGRAPHY

Allen, Arthur. *Vaccine: The Controversial Story of Medicine's Greatest Life-saver*. New York: Norton, 2007.

Angell, Marcia. *Science on Trial: The Clash of Medical Evidence and the Law in the Breast Implant Case*. New York: Norton, 1996.

Angier, Natalie. *The Canon: A Whirligig Tour of the Beautiful Basics of Science*. New York: Houghton Mifflin, 2007.

Boyce, Tammy. *Health, Risk and News: The MMR Vaccine and the Media*. Oxford: Lang, 2007.

Brandt, Allan M. *The Cigarette Century: The Rise, Fall, and Deadly Persistence of the Product That Defined America*. New York: Basic Books, 2007.

Fitzpatrick, Michael. *MMR and Autism: What Parents Need to Know*. London: Routledge, 2004.

Fumento, Michael. *Science Under Siege: How the Environmental Misinformation Campaign Is Affecting Our Lives*. New York: Morrow, 1996.

Green, Michael D. *Bendectin and Birth Defects: The Challenges of Mass Toxic Substances Litigation*. Philadelphia: University of Pennsylvania Press, 1996.

Grinker, Roy Richard. *Unstrange Minds: Remapping the World of Autism*. New York: Basic Books, 2007.

Heath, Chip, and Dan Heath. *Made to Stick: Why Some Ideas Survive and Others Die*. New York: Random House, 2007.

Horton, Richard. *MMR Science and Fiction: Exploring the Vaccine Crisis*. London: Granta Books, 2004.

——. *Second Opinion: Doctors, Diseases and Decisions in Modern Medicine*. London: Granta Books, 2003.

Huber, Peter. *Galileo's Revenge: Junk Science in the Courtroom*. New York: Basic Books, 1991.

Institute of Medicine. *Immunization Safety Review: Measles-Mumps-Rubella Vaccine and Autism*. Washington, DC: National Academies Press, 2001.

——. *Immunization Safety Review: Thimerosal-Containing Vaccines and Neurodevelopmental Disorders*. Washington, DC: National Academies Press, 2001.

——. *Immunization Safety Review: Vaccines and Autism*. Washington, DC: National Academies Press, 2004.

Kirby, David. *Evidence of Harm: Mercury in Vaccines and the Autism Epidemic: A Medical Controversy*. New York: St. Martin's Press, 2005.

Milloy, Steven J. *Junk Science Judo: Self-Defense Against Health Scares and Scams*. Washington, DC: Cato Institute Press, 2001.

Milloy, Steven J., and Michael Gough. *Silencing Science*. Washington, DC: Cato Institute Press, 1998.

Moldin, Steven O., and John L. R. Rubenstein, eds. *Understanding Autism: From Basic Neuroscience to Treatment*. Boca Raton, FL: Taylor and Francis, 2006.

Offit, Paul. *The Cutter Incident: How America's First Polio Vaccine Led to the Growing Vaccine Crisis*. New Haven, CT: Yale University Press, 2005.

——. *Vaccinated: One Man's Quest to Defeat the World's Deadliest Diseases*. New York: Smithsonian Books, 2007.

Park, Robert. *Voodoo Science: The Road from Foolishness to Fraud*. Oxford: Oxford University Press, 2000.

Plotkin, Stanley A., Walter A. Orenstein, and Paul A. Offit, eds. *Vaccines*, 5th ed. London: Saunders Elsevier, 2008.

Schreibman, Laura. *The Science and Fiction of Autism*. Cambridge, MA: Harvard University Press, 2005.

Shermer, Michael. *Why People Believe Weird Things: Pseudoscience, Superstition, and Other Confusions of Our Time*. New York: Holt, 1997.

Taubes, G. *Bad Science: The Short Life and Weird Times of Cold Fusion*. New York: Random House, 1993.

U.S. Congress. House of Representatives. Committee on Government Reform. *Autism: Present Challenges, Future Needs: Why the Increased Rates?* Hearing before the Committee on Government Reform, 106th Congress, Second session, April 6, 2000. Washington, DC: U.S. Government Printing Office, 2001.

ACKNOWLEDGMENTS

I wish to thank Patrick Fitzgerald, publisher for the life sciences, Columbia University Press, for his wisdom and guidance; Andrew Zack for his unfailing belief in and support of this project; Bojana Ristich for her patient lessons on style and form; Rachel Klayman for discussions that led to chapter 11; Erica Johnson for her research assistance; and Jeffrey Bergelson, Bonnie Brier, Alan Cohen, Peggy Flynn, Jason Kim, Bonnie Offit, Carl Offit, Jason Schwartz, Kirsten Thistle, Michael Yudell, and Theo Zaoutis for their careful reading of the manuscript and helpful suggestions and criticisms.

I would also like to thank Jon Abramson, Arthur Allen, Curtis Allen, Carol Baker, Nathan Blum, Arthur Caplan, Camille Clark, Harvey Fineberg, Meg Fisher, Michael Fitzpatrick, Gary Freed, Anne Gershon, Peter Hotez, Susan Levy, Marie McCormick, John Modlin, Charlotte Moser, Martin Myers, Walter Olson, Walter Orenstein, Sarah Parker, Georges Peter, Diane Peterson, Larry Pickering, Amy Pisani, Lisa Randall, Lance Rodewald, David Salisbury, Anne Schuchat, Robert Schultz, Kathleen Seidel, Kimberly Strassel, Fred Volkmar, Deborah Wexler, Melinda Wharton, and Jon Yewdell for their recollections of the vaccines-autism controversy as well as their expertise.

INDEX

Note: Page numbers in *italics* indicate illustrations.